Reactivity and Structure
Concepts in Organic Chemistry

Volume 2

Editors:

Klaus Hafner   Jean-Marie Lehn
Charles W. Rees   P. von R. Schleyer
Barry M. Trost   Rudolf Zahradník

Kenichi Fukui

Theory of Orientation and Stereoselection

With 72 Figures

Springer-Verlag
Berlin Heidelberg New York 1975

Kenichi Fukui

Kyoto University, Dept. of Hydrocarbon Chemistry, Kyoto, Japan

ISBN 3-540 07426-0 Springer-Verlag Berlin Heidelberg New York
ISBN 0-387 07426-0 Springer-Verlag New York Heidelberg Berlin

Parts of this book have been published in Topics in Current Chemistry, Vol. 15 (1970)

Library of Congress Cataloging in Publication Data
Fukui, Kenichi, 1918-   Theory of orientation and stereoselection. (Reactivity and structure; v. 2)
Bibliography: p. Includes index. 1. Chemical reaction, Conditions and laws of.
2. Stereochemistry. 3. Chemistry, Physical organic. I. Title. II. Series.   QD501.F92   547'.1'223
75-25597
Printed in Germany.
Typesetting and printing: Hans Meister KG, Kassel, binding: Konrad Triltsch, Würzburg

Preface

Many organic chemists will agree with me that the old "electronic theory" has for a long time been inadequate for the interpretation of various new findings in chemistry, particularly for those of reactivity. Considering the outstanding progress which has been made during the past 20 years in the interpretation of these facts, aided by the molecular orbital theory, the time has finally come for a new book showing what is within and what is beyond the reach of quantum-chemical methods.

It was therefore highly suitable that Dr. F. L. Boschke of the Springer-Verlag suggested to me to make a contribution to a volume in the series "Topics in Current Chemistry" in February 1969. The article was published as Vol. 15, No 1 in June 1970. This new book is an expanded version of the article written in 1970.

In this present volume several of the most up-to-date findings which have been gained in organic chemistry since then have been added. It is highly probable that a certain "theoretical" design in the experimentalists' mind may have been the reason for these developments, whether they themselves are aware of it or not. Theory produces new experimental ideas and conversely, a host of experimental data add another vista to new theories.

Due to the mutual beneficial effect of theory and experiment this book will always retain its value, although the quantum-chemical approach to the theory of reactivity is, of course, still in the developmental stage.

It is my sincere hope that graduates and young research chemists, in both the theoretical and experimental fields will find this book useful and thereby become acquainted with the quantum-chemical way of thinking in which the concept of the "orbital" of an electron serves as a good explanation within the chemical terminology.

I extend my special thanks to Dr. F. L. Boschke and Springer-Verlag for the planning and production of this book.

KENICHI FUKUI

Kyoto, Japan, May 1975

Contents

1. Molecular Orbitals

Many chemical problems can be discussed by way of a knowledge of the electronic state of molecules. The electronic state of a molecular system becomes known if we solve the electronic *Schrödinger equation*, which can be separated from the time-independent, nonrelativistic Schrödinger equation for the whole molecule by the use of the Born-Oppenheimer approximation [1]. In this approximation, the electrons are considered to move in the field of momentarily fixed nuclei. The nuclear configuration provides the parameters in the Schrödinger equation.

The nonrelativistic, electronic *Schrödinger Hamiltonian operator*, designated as H, is represented by

$$H = \sum_{i=1}^{N} \left(-\frac{h^2}{8\pi^2 m} \Delta_i - \sum_a \frac{Z_a e^2}{r_{ia}} \right) + \sum_{\substack{i,j=1 \\ (i<j)}}^{N} \frac{e^2}{r_{ij}} + \sum_{\substack{a,b \\ (a<b)}} \frac{Z_a Z_b e^2}{r_{ab}} \qquad (1.1)$$

in which

N is the number of electrons,

Δ_i is the Laplacian operator for electron i,

$Z_a e$ is the positive charge of nucleus, a, and

r_{ij}, r_{ia}, and r_{ab} are the distances between electrons i and j, nucleus a and electron i, and nuclei a and b,

respectively: e and m are the charge and the mass of an electron: h is the Planck constant.

The eigenstate of the operator H may be described in terms of $4\,N$ electron coordinates,

$$x_i, y_i, z_i, \quad \text{and} \quad \xi_i \ (i = 1, 2, \text{-----} \ N),$$

where the first three are the Cartesian coordinates and the last one is the spin coordinate. The *wave function*, Ψ, of an eigenstate of H is therefore

1

represented by $\Psi(12 ----- N)$ in which $i\,(i=1, 2, ----- N)$ stands for the set of coordinates (x_i, y_i, z_i, ξ_i).

From the well-known statistical requirement for an assembly of Fermi particles, $\Psi(12 ----- N)$ is subject to a limitation in its form of anti-symmetric character with respect to electron exchange. In addition to this, we have to note that an eigenstate of H can be specified also by the *eigenvalues* of $\mathbf{S}^2$ and S_z, where $\mathbf{S}$ is the *total electronic spin angular momentum vector*. In this way, we are able to obtain information about the general form which should be satisfied by the simultaneous eigenfunction of H, $\mathbf{S}^2$, and S_z. Let such a function be denoted by Ψ_{SM_S} in which S and M_S specify the eigenvalues of $\mathbf{S}^2$ and S_z, respectively. In this way, the form which must be taken by an antisymmetric spin-eigenstate N-electron wave function can be derived.

For instance, as is well known, the general form of wave functions with $N=2$, $S=0$, $M_S=0$ is

$$\{\psi(12) + \psi(21)\}\{\alpha(1)\,\beta(2) - \beta(1)\,\alpha(2)\} \tag{1.2}$$

where $\psi(12)$ is an arbitrary two-electron spatial function, and α and β are the usual spin functions. If an "exact" eigenfunction of H for a two-electron system were obtained, it would naturally be of this form.

Such a "general" form of wave function is easily written explicitly for each set of values of N, S, and M_S. Any appropriate form of approximate wave functions, like determinantal functions composed of one-electron functions ("molecular spin orbitals"), the "bond eigenfunctions" used in the valence bond approach, and so on, is shown to fulfil this requirement.

Some of these approximate forms of wave function possess a character of particular theoretical interest. One such is the *"uni-configurational"* *wave function*. This implies an appropriate linear combination of anti-symmetrized products of molecular spin orbitals in which all antisymmetrized products belong to the same "electron configuration". The electron configuration of an antisymmetrized product is defined as the set of N spatial parts appearing in the product of spin orbitals. For instance, a uni-configurational wave function with $N=2$, $S=0$, $M_S=0$ is expressed as

$$(i\bar{j}) - (\bar{i}j) \tag{1.3}$$

where

$$(i\bar{j}) \equiv \begin{vmatrix} i(1)\,\alpha(1)\ j(1)\,\beta(1) \\ i(2)\,\alpha(2)\ j(2)\,\beta(2) \end{vmatrix} \quad \text{etc.,}$$

and the set $[ij]$ stands for the electron configuration. The spatial part of a spin orbital is often called simply an *"orbital"*. The orbital which appears only once in an electron configuration is said to be "singly occupied", and that appearing twice "doubly occupied".

The general form of such uni-configurational wave functions can be obtained for any set of N, S, and M_S. It is easy to see that such a form of wave functions duly satisfies the general requirement mentioned above, as in Eq. (1.2).

Some uni-configurational wave functions consist of only one determinant. This is called a *single-determinant wave function*. A single-determinant can be a spin-eigenstate wave function only if the eigenfunctions possess the values of

$$S = |M_S| = \tfrac{1}{2}(N - 2\nu)$$

where ν is the number of doubly occupied orbitals in the determinant. Thus

[*case A*] open-shell wave functions with maximum multiplicity $(\nu = 0, S = |M_S| = N/2)$,

[*case B*] closed-shell wave functions $(\nu = N/2, S = |M_S| = 0)$, and

[*case C*] wave functions with a closed-shell structure of ν doubly occupied orbitals with additional open-shell structure of $S = |M_S| = \tfrac{1}{2}(N - 2\nu)$ belong to this category. Any other uni-configurational wave functions consist of more than one determinant.

We can discuss the "best" uni-configurational wave function by the usual variational method of the *Hartree-Fock* type. This means making a search for the function Ψ which minimizes the quantity

$$\int \Psi^* H \, \Psi \, d\tau / \int \Psi^* \, \Psi \, d\tau . \qquad (1.4)$$

If an excited state is concerned, this is done under the restriction that the function should be orthogonal to all of the lower-energy states. We may specify these as the *"uni-configurational Hartree-Fock wave functions"*. The "best" orbitals constructing the determinants in these wave functions are in general not orthogonal to each other.

In [*case A*] and [*case B*] mentioned above, the "best" wave function thus obtained is of particular practical importance. The set of N orbitals appearing in these functions is in general definitely determined, except for an arbitrary numerical factor of which the absolute value is unity, as being mutually orthogonal and having a definite "orbital energy" [cf.

3

Eq. (3.15)]. The concept of "electron occupation" of orbitals is thus unequivocal in these cases. The best orbitals in these cases are called "Hartree-Fock orbitals"[2,3].

The wave function of [*case A*] is in general written in the form

$$\frac{1}{\sqrt{N!}} \begin{vmatrix} \phi_1(1) & \phi_2(1) & \text{------} & \phi_N(1) \\ \phi_1(2) & \phi_2(2) & \text{-------} & \phi_N(2) \\ \text{------------------------} \\ \phi_1(N) & \phi_2(N) & \text{----} & \phi_N(N) \end{vmatrix} \sigma^{(S)}(1, 2, \text{-----} N) \qquad (1.5)$$

where $\phi_i(k)$ is the ith orbital occupied by the *kth* electron and $\sigma^{(S)}(1, 2, \text{-----} N)$ is the totally symmetric N-electron spin function.

The wave function of [*case B*] with $N=2$ can be written as

$$\phi_1(1)\, \phi_1(2)\, \frac{1}{\sqrt{2}}\, \{\alpha(1)\, \beta(2) - \beta(1)\, \alpha(2)\} \qquad (1.6)$$

The closed-shell wave functions with $N>2$ can no longer be separated into spatial and spin parts, but are expressed in the following form:

$$\frac{1}{\sqrt{(2\nu)!}} \begin{vmatrix} \phi_1(1)\,\alpha(1)\, \phi_1(1)\,\beta(1)\, \phi_2(1)\,\alpha(1)\, \phi_2(1)\,\beta(1) \text{---} \phi_\nu(1)\,\alpha(1)\, \phi_\nu(1)\,\beta(1) \\ \phi_1(2)\,\alpha(2) \text{------------------------------} \phi_\nu(2)\,\beta(2) \\ | \\ \text{--} \\ | \\ \phi_1(N)\,\alpha(N) \text{----------------------------} \phi_\nu(N)\,\beta(N) \end{vmatrix} \qquad (1.7)$$

Such a determinantal form of wave function is often called the Slater determinant.

Thus, we have the N-electron wave function with separated spatial and spin parts only in the cases of two-electron singlet states and N-electron $(N+1)$-plet states. The Hartree-Fock orbitals are defined as those functions ϕ_i which make the wave functions (1.5), (1.6), and (1.7) best. The usual variation technique leads to the N(case A) or ν(case B) simultaneous differential equations which have to be satisfied by ϕ_i $(i=1, 2, \text{---} N$ in case A, and $i=1, 2, \text{---} \nu$ in case B). These equations are called the Hartree-Fock equations. The Hartree-Fock orbitals are obtained by solving these differential equations simultaneously.

Besides the occupied orbitals, these equations possess solutions corresponding to actually unoccupied, virtual orbitals. Some of them happen to possess negative energies (corresponding to "bound one-electron states"), whereas the others have nonnegative energies. The Hartree-Fock unoccupied orbital, rather than its realistic physical meaning, is important in the sense that it is used in *constructing excited-state wave functions* and plays a significant role in the theory of chemical interactions (Chap. 3). It is to be remarked that the mathematical means suitable for describing the unoccupied orbitals are not always the same as those representing the occupied orbitals with tolerable approximation.

The Hartree-Fock equations for the *hydrogen molecule* have been solved by Kolos and Roothaan[4], by obtaining the binding energy value of 3.63 eV for the ground state, which is ca. 1.1 eV smaller than the exact theoretical value [4,5]. This difference corresponds to the correlation error. The Hartree-Fock orbital energies of other *homonuclear diatomic molecules*, C_2, N_2, O_2 and F_2, have been obtained by Buenker *et al.* [6]. A review has been given by Wahl *et al.* [7] with illustrative orbital maps for the F_2, NaF, and N_2 molecules. Also calculations have been made with respect to simple *hydrocarbons* such as CH_4, C_2H_6, C_2H_4, and C_2H_2 [6,8,9].

The Hartree-Fock orbitals are expanded in an infinite series of known basis functions. For instance, in diatomic molecules, certain two-center functions of elliptic coordinates are employed. In practice, a limited number of appropriate atomic orbitals (AO) is adopted as the basis. Such an approach has been developed by Roothaan [10]. In this case the Hartree-Fock differential equations are replaced by a *set of nonlinear simultaneous equations* in which the limited number of AO coefficients in the linear combinations are unknown variables. The orbital energies and the AO coefficients are obtained by solving the Fock-Roothaan secular equations by an iterative method. This is the procedure of the Roothaan LCAO (linear-combination-of-atomic-orbitals) SCF (self-consistent-field) method.

The basis AO adopted may be Slater-type orbital (STO) [11], Gaussian-type orbital (GTO) [12], and Hartree-Fock AO [13], Löwdin's orthogonalized AO [14], and so on. In many cases the Slater AO's for the valence-shell electrons are taken. Clementi has extended the basis beyond the valence shells [15]. Frequently, the exponents of Slater AO's are optimized. Clementi has also adopted two different variable exponents for "one" Slater AO [15].

Even an exact Hartree-Fock calculation cannot be exempt from the correlation error. A practical method of evaluation has been proposed by Hollister and Sinanoğlu [16]. An *LCAO SCF method* has been applied to the calculation of the heat of various simple reactions by Snyder and Basch [17]. They have evaluated the correlation error by the method of Hollister and Sinanoğlu [16].

In the cases other than [*case A*] and [*case B*], so called *"open-shell"* *SCF methods* are employed. The orbital concept becomes not quite certain. The methods are divided into classes which are "restricted" [18] and "unrestricted" [19] Hartree-Fock procedures. In the latter case the wave function obtained is no longer a spin eigenfunction.

The Hartree-Fock method is modified by mixing some important valence electron configurations with the ground-state one [20]. This is called the *OVC* (*optimized valence configurations*) *method*.

Such a wave function is represented by a linear combination of wave functions for more than one electron configuration, and is called a *"multi-configurational"* *wave function*. The consideration of more than one configuration can reduce the correlation error. Such an approach is referred to as the *method of "configuration interaction* (CI)".

Some useful, conventional SCF methods have been proposed by Pople [21] and by Kon [22] using the semiempirical calculation of Pariser and Parr [23] with regard to the π electrons of planar conjugated molecules.

Yonezawa *et al.* [24] have developed an SCF method taking into account all valence electrons with all overlap integrals included. They have made calculations with respect to several simple molecules, such as

CH_4, C_2H_6, C_2H_4, C_2H_2, CO, CO_2, H_2O, H_2CO, CH_3OH, HCN, and NH_3 [24];

larger molecules like butadiene, acrolein, and glyoxal [25]; several alkyl radicals of $C_1 \sim C_4$ [26]; and aza-heterocycles [27]. This method gives reasonable theoretical values for transition energies, ionization potentials, dipole moments, and chemical reactivities of these molecules.

A method which is similar to the Pariser-Parr-Pople method for the π electron system and is applicable to common, saturated molecules has been proposed by Pople [28a]. This method is called the *CNDO* (*complete neglect of differential overlap*) *SCF calculation*. Some further modifications have been made — *INDO* (*Intermediate Neglect of Differential Overlap*) [28b] and *MINDO* (*Modified INDO*) [28c] methods. Katagiri and Sandorfy [29] and Imamura *et al.* [30] have used hybridized orbitals as basis of the Pariser-Parr-Pople type semiempirical SCF calculation.

Other approximate, more empirical methods are the extended *Hückel* [31] *and hybrid-based Hückel* [32,33] *approaches*. In these methods the electron repulsion is not taken into account explicitly. These are extensions of the early Hückel molecular orbitals [34] which have successfully been used in the π electron system of planar molecules. On account of the simplest feature of calculation, the Hückel method has made possible the first quantum mechanical interpretation of the classical electronic theory of organic chemistry and has given a reasonable explanation for the chemical reactivity of sizable conjugated molecules.

Development of quantum-chemical calculation of the character antipodal to this simple and empirical one — *nonempirical approaches* in which no empirical data are employed — has been promoted by the recent progress of high-speed computers. The point of success was the usage of GTO which facilitates the computation of multicentre integrals. The expansion technique of STO into nGTO's (STO–nG) [35] in the least square fitting contributed much. The variation of the total energy of a system with n was investigated [36]. The calculation of molecular geometry [37], heat of reaction [38], activation energy [39], and potential barrier to internal rotation [40], was made. Clementi obtained the energy of hydrogen bonding in guanine-cytosine base pair by using 334 GTO's[41]. An excellent review for nonempirical calculations is available [42].

2. Chemical Reactivity Theory

From 1933 [43], several theoretical approaches to the problem of the chemical reactivity of planar conjugated molecules began to appear, mainly by the Hückel molecular orbital theory. These were roughly divided into two groups [44]. The one was called the *"static approach"* [43,45–48], and the other, the *"localization approach"* [49,50]. In 1952, another method which was referred to as the *"frontier-electron method"* was proposed [51] and was conventionally grouped [52] together with other related methods [53,54] as the *"delocalization approach"*.

The first paper of the frontier-electron theory pointed out that the *electrophilic aromatic substitution* in aromatic hydrocarbons should take place at the position of the greatest density of electrons in the *highest occupied* (HO) molecular orbital (MO). The second paper disclosed that the nucleophilic replacement should occur at the carbon atom where the *lowest unoccupied* (LU) MO exhibited the maximum density of extension. These particular MO's were called "frontier MO's". In homolytic replacements, both HO and LU were shown to serve as the frontier MO's. In these papers the "partial" density of $2\,p\pi$ electron, in the HO (or LU) MO, at a certain carbon atom was simply interpreted by the square of the atomic orbital (AO) coefficient in these particular MO's which were represented by a linear combination (LC) of $2\,p\pi$ AO's in the frame of the Hückel approximation. These partial densities were named "frontier-electron densities".

The explanation of these findings was at that time never self-evident. In contrast to the other reactivity theories, which then existed and had already been well-established theoretically, the infant frontier-electron theory was short of solid physical ground, having suggested a possibility of the involvement of a new principle relating to the nature of chemical reactions.

In the same year as that of the proposal of the frontier-electron theory, the *theory of charge-transfer force* was developed by Mulliken with regard to the molecular complex formation between an electron donor and an acceptor [55]. In this connection he proposed the "overlap and orientation" principle [56] in which only the overlap interaction between the HO MO of the donor and the LU MO of the acceptor is considered.

The behaviour of the frontier electrons was also attributed to a certain type of electron delocalization between the reactant and the reagent [57]. A concept of *pseudo-π-orbital* was introduced by setting up a simplified model, and the electron delocalization between the π-electron system of aromatic nuclei and the pseudo-orbital was considered to be essential to aromatic substitutions. The pseudo-orbital was assumed to be built up out of the hydrogen atom AO attached to the carbon atom at the reaction center and the AO of the reagent species, and to be occupied by zero, one, and two electrons in electrophilic, radical, and nucleophilic reactions. A theoretical quantity called "superdelocalizability" was derived from this model. This quantity will be discussed in detail later in Chap. 6.

a) Reaction with an
   electrophilic reagent

b) Reaction with a
   radical reagent

c) Reaction with a
   nucleophilic reagent

π-System Pseudo-orbital   π-System Pseudo-orbital   π-System Pseudo-orbital

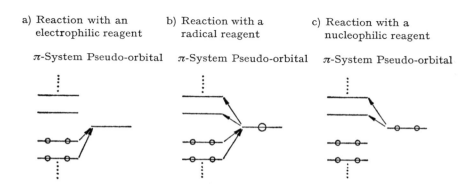

The frontier-electron density was used for discussing the reactivity within a molecule, while the superdelocalizability was employed in comparing the reactivity of different molecules [52]. Afterwards, the applicability of the frontier-electron theory was extended to saturated compounds [58]. The new theoretical quantity "delocalizability" was introduced for discussing the reactivity of saturated molecules [58]. These indices satisfactorily reflected experimental results of various chemical reactions. In addition to this, the conspicuous behavior of HO and LU in determining the steric course of organic reactions was disclosed [52,59].

All of these facts make one believe that the distinction of particular MO's, the frontier orbitals, from the others has a good reason which arises from the general principle governing the nature of chemical reactions. It is useful in this connection to analyze first the interaction energy of two reacting species in general [60]. The energy is divided into several terms so that one can understand what kind of interaction energy is really important in chemical reactions.

3. Interaction of Two Reacting Species

Two isolated reactant molecules in the closed-shell ground state are designated as A and B, whose electronic energies are W_{AO} and W_{BO}, respectively. Here the term closed-shell implies the structure of a molecule with doubly occupied MO's only. The lowest total energy of the two mutually interacting systems is denoted by W. Then, the interaction energy is defined by

$$\Delta W = W - (W_{AO} + W_{BO}) \tag{3.1}$$

All the energy values are calculated by the Born-Oppenheimer approximation with respect to a fixed nuclear configuration. The most stable configurations of interacting systems are obviously different from the respective isolated systems. However, the nuclear configuration change is tentatively left untouched in order to disclose the constitution of interaction energy at the beginning of the theory. Namely, W is the energy of a system composed of A and B approaching each other without deformation, satisfying the Schrödinger equation for the combined system

$$H\Psi = W\Psi \tag{3.2}$$

in which the Hamiltonian operator H is represented by

$$H = \sum_{\lambda} H(\lambda) + \sum_{\lambda < \lambda'} \frac{e^2}{r_{\lambda\lambda'}} + \sum_{\gamma < \gamma'} \frac{Z_\gamma Z_{\gamma'} e^2}{R_{\gamma\gamma'}} \tag{3.3}$$

$$H(\lambda) = - \frac{h^2}{8\pi^2 m} \Delta(\lambda) + V(\lambda) \tag{3.4}$$

$$V(\lambda) = V_A(\lambda) + V_B(\lambda) \tag{3.5}$$

$$V_A(\lambda) = - \sum_{a} \frac{Z_a e^2}{r_{\lambda a}}, \qquad V_B(\lambda) = - \sum_{\beta} \frac{Z_\beta e^2}{r_{\lambda\beta}} \tag{3.6}$$

10

$H(\lambda)$ is the one-electron Hamiltonian operator of the electron λ

Z_α, Z_β, and Z_γ are the positive charge numbers of the nuclei α, β, and γ, belonging to molecule A, molecule B, and the combined system, AB

$r_{\lambda\lambda'}$ is the distance between the two electrons λ and λ':

$r_{\lambda\alpha}$ is the distance of the electron λ from the fixed nucleus α

$R_{\alpha\alpha'}$ is the distance between the fixed nuclei α and α'

$\Delta(\lambda)$ is the Laplacian operator for the electron λ

$V_A(\lambda)$ and $V_B(\lambda)$ are the potential energies of the electron λ due to the nuclei belonging to molecules A and B, respectively.

To compose the wave function Ψ for the combined system A ————— B, an attempt is made to employ the MO's of the isolated reactant molecules A and B. The unperturbed normalized wave functions of A and B are represented in terms of the Slater determinants composed of ortho-normal (mutually orthogonal (cf. Chap. 1) and normalized) spin orbitals. The spin orbitals are assumed to have the spatial parts which are made SCF MO with respect to the *ground state* of each isolated molecule, A or B, in the Hartree-Fock sense (Chap. 1). To make an approximate excited-state wave function of an isolated system, the Hartree-Fock unoccupied MO's mentioned in Chap. 1 which are associated with the Hartree-Fock equation for the ground state are employed in constructing the Slater determinant. In this way, all of the MO's which are used in the wave function for the combined system A ————— B are defined definitely with regard to a given nuclear configuration in each isolated system.

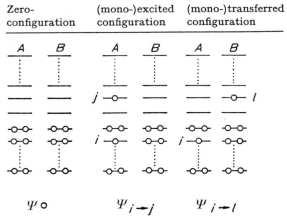

Fig. 3.1. The electron configuration of the combined system A ————— B

11

The wave function Ψ for the combined system A ------ B is represented by a multi-configurational one which is a linear combination of the spin-eigenstate determinantal functions composed of the above-defined spin orbitals, which are antisymmetrized with respect to all electrons of the whole system. These determinantal functions correspond to the *electron configurations* illustrated below.

The zero-configuration corresponds to the combined system in which A and B interact in their ground state. In an excited configuration either A or B (or both) is in an excited state. The transferred configuration is one in which one (or more) electron is transferred from an MO of one system to an MO of the other. The MO's occupied and unoccupied in the ground state are discerned by the following notation:

system A　　　system B

unoccupied : $a_j\ a_{j'}$ ......{ ___b_l　b_l '.....{ ___

occupied : $a_i\ a_{i'}$ ......{ b_k　$b_{k'}$.....{

The wave function is thus represented by

$$\Psi = C_0\,\Psi_0 + (\underset{p}{\Sigma} + \underset{p}{\Sigma} + \underset{p}{\Sigma} + \underset{p}{\Sigma} + \underset{p}{\Sigma} + ---) C_p\,\Psi_p \qquad (3.7)$$

monoex.　monotr.　diex.　monoex.–monotr.　ditr.

where suffix 0 implies the zero-configuration, *ex.* and *tr.* signify "*excited*" and "*transferred*", respectively, Ψ_p represents one of the wave functions $\Psi_{i\to j}$, $\Psi_{i\to l}$, etc. corresponding to the above-depicted electron configurations, or highly excited or transferred ones, and C_0 and C_p are coefficients which are to be determined so as to minimize the total energy of the combined system A ------ B.

An approach such as this belongs to the method of configuration interaction (CI) mentioned in Chap. 1. It is sufficient to cite a simple example to illustrate the usefulness of such CI treatments. It is well known that the Weinbaum wave function [61] for the hydrogen molecule gives a better result than the Hartree-Fock calculation, notwithstanding the simplest form as follows:

$$\Phi(1,2) = a\,\{\chi_A(1)\,\chi_B(2) + \chi_B(1)\,\chi_A(2)\} + b\,\{\chi_A(1)\,\chi_A(2) + \chi_B(1)\,\chi_B(2)\}$$

where χ_A and χ_B are the $1s$ AO of the hydrogen atoms A and B with the effective nuclear charge larger than unity $(= 1.193)$. This implies that the AO is "shrunken". Mulliken [62] has shown that the $1s$ function with orbital exponent 1.2 can be expanded in terms of ns functions with orbital exponent unity:

$$(1s)(\zeta = 1.2) = 0.9875(1s)(\zeta = 1) - 0.0925(2s)(\zeta = 1)$$
$$+ 0.0433(3s)(\zeta = 1) - - - - - -$$

On substituting such an expansion into the wave function formula, it becomes evident that this function consists of the following terms:

which corresponds obviously to Eq. (3.7). The success of Weinbaum's treatment may be attributed to the CI nature of that treatment.

In view of the impracticability of the Hartree-Fock calculation for common molecules, the LCAO MO spatial functions may be used in place of Hartree-Fock ones. The MO's a and b are given by

$$a(1) = \sum_t c_t\, t(1) \qquad \text{for molecule } A$$

$$b(1) = \sum_u c_u\, u(1) \qquad \text{for molecule } B$$

(3.8)

13

where 1 implies the coordinates of the electron 1, and t and u are the AO's belonging to the nuclei of A and B, respectively. The coefficients c_t and c_u are chosen so that $a(1)$ and $b(1)$ become Roothaan-type SCF MO's [10] for the ground state of each isolated system. The AO's $t(1)$ and $u(1)$ may usually be taken to be real Slater-type AO's, for example.

The total energy W in Eq. (3.2) is obtained by solving the usual secular equation [52] as

$$W = H_{0,0} - \left(\overset{\text{monoex.}}{\underset{p}{\sum}} + \overset{\text{monotr.}}{\underset{p}{\sum}} + \overset{\text{diex.}}{\underset{p}{\sum}} + \overset{\substack{\text{monoex.} - \\ \text{monotr.}}}{\underset{p}{\sum}} + \overset{\text{ditr.}}{\underset{p}{\sum}} + \, --- \right)$$

(3.9)

$$\frac{|H_{0,p} - S_{0,p} H_{0,0}|^2}{H_{p,p} - H_{0,0}} + \, ---$$

where

$$H_{p,q} = \int \Psi_p^* H \Psi_q \, d\tau \quad \text{and} \quad S_{p,q} = \int \Psi_p^* \Psi_q \, d\tau$$

and the wave function Ψ in Eq. (3.7) is simultaneously determined.

Consider the case where the interaction between the molecules A and B is not yet very strong. The magnitude of $H_{0,p}$ is almost linear with $S_{0,p}$, so that the second-order term in Eq. (3.9) is proportional to the square of $S_{0,p}$. The order of magnitude of $S_{0,p}$ is equal to the γth power of an overlap integral s_{ab} of an MO a of the molecule A and an MO b of the molecule B, where γ is the minimum number of electron transfers between A and B required to shift the electron configuration from 0 to p. Therefore, the terms from monotransferred configurations in Eq. (3.9) have magnitudes of the order of s_{ab}^2, while the monoex. and the ditr. terms are of s_{ab}^4, and the monoex.-monotr. term s_{ab}^6, the diex. term s_{ab}^8, and so on. If the interaction is weak and s_{ab} is small, the mono-transferred terms are important in comparison with the others.

There are some additional reasons which make the contribution of *monotransferred terms* uniquely important. As assumed before, the MO's used are the Hartree-Fock or other SCF ones so that the values of $H_{0,p}$ of monoex. terms are small, since the Brillouin theorem [63] requires that the matrix element between the ground state and a monoexcited state in the Hartree-Fock approach should vanish in an isolated molecule. In addition to this, the denominator of the second-order term

14

$(H_{p,q} - H_{0,0})$ in Eq. (3.9) can usually not be small in excited configuration terms, whereas in transferred configuration terms it can be. Even a first-order term of the form

$$- | H_{0,p} - S_{0,p} H_{0,0} | \tag{3.10}$$

appears in place of the second-order term

$$- \frac{| H_{0,p} - S_{0,p} H_{0,0} |^2}{H_{p,p} - H_{0,0}}$$

when $H_{p,p}$ is approximately equal to $H_{0,0}$, that is, in a "degenerate" case. From these considerations, the following approximate formula is obtained:

$$W \cong H_{0,0} - \sum_{p}^{\text{monoex.}} \frac{| H_{0,p} - S_{0,p} H_{0,0} |^2}{H_{p,p} - H_{0,0}} \tag{3.11}$$

The interaction energy, ΔW, in Eq. (3.1) is in this way converted into the form

$$\Delta W \cong \varepsilon_Q + \varepsilon_K - D \tag{3.12} \text{ 64)}$$

where ε_Q is the Coulomb interaction term represented by

$$\varepsilon_Q \cong \sum_{a} \sum_{\beta} e^2 \frac{(Z_a - N_a)(Z_\beta - N_\beta)}{R_{a\beta}} \tag{3.13}$$

by the use of Mulliken's approximation [65], in which N_a is the population of electrons, so that $e(Z_\alpha - N_\alpha)$ is the net plus charge, of the atom α, ε_K is the *exchange interaction term*, and D is the stabilization energy due to the electron delocalization interaction, which is written in the following form

$$D = \sum_{i}^{\text{occ}} \sum_{l}^{\text{uno}} \frac{|H_{0,i \to l} - S_{0,i \to l} H_{0,0}|^2}{H_{i \to l, i \to l} - H_{0,0}} + \sum_{k}^{\text{occ}} \sum_{j}^{\text{uno}} \frac{|H_{0, k \to j} - S_{0, k \to j} H_{0,0}|^2}{H_{k \to j, k \to j} - H_{0,0}} \tag{3.14}$$

15

where $\overset{occ}{\sum}$ and $\overset{uno}{\sum}$ imply the summation covering the occupied and un-occupied MO's, respectively.

The form of Eq. (3.13) indicates that this term is the sum of Coulomb potentials arising from the net charge of each atom of molecule A and that of each atom of molecule B. Therefore, ε_Q is significant in the interaction of polar molecules, causing a long-range force.

The *exchange interaction term*, ε_K, is important in the short range, being as usual repulsive in the interaction of closed-shell molecules, although it behaves as attractive in the singlet interaction of two odd-electron systems. Suppose that the overlapping of MO's of A and B takes place appreciably only between one AO, say r, of A and one AO, say r', of B. Such a mode of interaction may be called single-site overlapping, and is nearly realized in the aromatic substitution by a reagent with essentially one AO. In such cases the exchange interaction terms vary with the square of the overlap integral $s_{rr'}$, so that they are less important than the Coulomb term, at least at the initial stage of interaction of two closed-shell molecules.

The term D of Eq. (3.14) is called the *delocalization stabilization*, which is usually positive. This term comes from the electron delocalization between the molecules A and B. The physical meaning of the denominator of each term in the right side of Eq. (3.14) can be discussed in relation to the Koopmans theorem [66]

$$\varepsilon_i = -I_i \tag{3.15}$$

in which ε_i is the energy of the ith MO and I_i is the ionization potential with respect to the electron in the ith MO. From the result of calculation [52] it follows that

$$H_{i \to l, i \to l} - H_{0,0} = I_{Ai}^{(B)} - E_{Bl}^{(A-i)} \tag{3.16}$$

$$= I_{Ai}^{(B+l)} - E_{Bl}^{(A)}$$

where I_{Ai} is the ionization potential of A with respect to the ith MO and E_{Bl} is the electron affinity of B with respect to the lth MO, and $I_{Ai}^{(B)}$ signifies the I_{Ai} value in the case of the approach of molecule B, $E_{Bl}^{(A)}$ is the value of E_{Bl} with the approach of molecule A, $I_{Ai}^{(B+l)}$ is the I_{Ai} in the approach of molecule B with an additional electron in the lth MO which is unoccupied in the ground state, and $E_{Bl}^{(A-i)}$ is the value of E_{Bl} in the case of the approach of molecule A in which one electron in the

16

*i*th MO is subtracted. The relation of Eq. (3.16) is schematically represented by the following figure:

$$H_{i-l,\,i-l} - H_{0,0}$$

$$= I_{Ai}^{(B)} - E_{Bl}^{(A-i)}$$

$$= I_{Ai}^{(B+l)} - E_{Bl}^{(A)}$$

With long intermolecular distance the integrals appearing in the numerator of each term in the right side of Eq. (3.14) can be rewritten as [60]

$$H_{0,\,i\to l} - S_{0,\,i\to l}\,H_{0,0} = 2\sum_{r} c_{r}^{(t)}\,c_{r'}^{(l)}\,\gamma_{rr'}^{(i)} \tag{3.17}$$

in which the multiple-site interaction between molecules A and B is assumed to take place through a paired overlapping of the rth AO of A and the r'th AO of B,

where

$$\gamma_{rr'}^{(i)} \cong -\int r(1)\left(\sum_{\beta}\frac{Z_{\beta}-N_{\beta}}{r_{1\beta}}\right)r'(1)\,dv(1) + s_{rr'}\sum_{a}\sum_{\beta}\frac{n_{a}^{(ii)}\,(Z_{\beta}-N_{\beta})}{R_{a\beta}} \tag{3.18}$$

and

$$n_{a}^{(ii)} = \sum_{t}\sum_{t'}^{(a)} c_{t}^{(t)}\,c_{t'}^{(t)}\,s_{tt'} \tag{3.19}$$

17

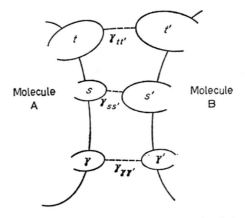

Fig. 3.2. The mode of multiple-site overlapping interaction

in which $c_t^{(i)}$ is the coefficient of the tth AO in the ith MO as in Eq. (3.8.) Eq. (3.18) indicates that the charge-transfer interaction in the initial stage is governed by the local net charge near the reaction center of the acceptor molecule.

In this way, the expression for the delocalization stabilization, D, is obtained as

$$D \sim 2 \left\{ \sum_i^{occ} \sum_l^{uno} \frac{(\sum_r c_r^{(i)} c_{r'}^{(l)} \gamma_{rr'}^{(i)})^2}{I_{Ai}^{(B)} - E_{Bl}^{(A-i)}} + \sum_k^{occ} \sum_j^{uno} \frac{(\sum_r c_r^{(j)} c_{r'}^{(k)} \gamma_{rr'}^{(k)})^2}{I_{Bk}^{(A)} - E_{Aj}^{(B-k)}} \right\} \quad (3.20)$$

This quantity represents the energy of the multiple-site electron delocalization interaction which will later play an important role in the theory of stereoselection. It is to be remarked that, although any MO may involve an arbitrary constant of which the absolute value is unity, the value of the numerator in each term of the right side of this equation is always definite.

One of the most important special cases is that of the single-site interaction between the rth AO of the reactant, A, and a reagent, B, which possesses only one AO designated as r'. In this case D is written as

$$D \sim \begin{cases} 2 \sum_i^{occ} \dfrac{c_r^{(i)2}}{\varepsilon_B - \varepsilon_{Ai}} \gamma_r^2 & \text{(the reagent orbital is unoccupied)} \quad (3.21a) \\[2em] 2 \sum_j^{uno} \dfrac{c_r^{(j)2}}{\varepsilon_{Aj} - \varepsilon_B} \gamma_r'^2 & \text{(the reagent orbital is occupied)} \quad (3.21b) \end{cases}$$

18

where

$$\gamma_r = c_r^{(l)}\, \gamma_{rr'}^{(l)},\; \gamma_r' = c_{r'}^{(k)}\, \gamma_{rr'}^{(k)}, \quad \varepsilon_{Ai} = -I_{Ai}^{(B)}, \quad \varepsilon_{Aj} = -E_{Aj}^{(B-k)},$$

and

$$\varepsilon_B = \begin{cases} -E_B^{(A-i)} & \text{(the reagent orbital is unoccupied)} & (3.22\,\text{a}) \\[2ex] -I_B^{(A)} & \text{(the reagent orbital is occupied)} & (3.22\,\text{b}) \end{cases}$$

The right sides of Eq. (3.21) can be employed as a measure of the chemical reactivity of both saturated and unsaturated compounds, which will be discussed in detail later.

The case of interaction between an *even-electron molecule A* and an *odd-electron molecule B* can be discussed in a similar manner. Eq. (3.20) is modified to be

$$D \sim 2 \left\{ \sum_i^{occ} \sum_l^{uno} \frac{(\sum_r c_r^{(i)} c_{r'}^{(l)} \gamma_{rr'}^{(i)})^2}{I_{Ai}^{(B)} - E_{Bl}^{(A-i)}} + \sum_k^{occ} \sum_j^{uno} \frac{(\sum_r c_r^{(j)} c_{r'}^{(k)} \gamma_{rr'}^{(k)})^2}{I_{Bk}^{(A)} - E_{Aj}^{(B-k)}} \right\}$$

$$+ \left\{ \sum_i^{occ} \frac{(\sum_r c_r^{(i)} c_{r'}^{(o')} \gamma_{rr'}^{(i)})^2}{I_{Ai}^{(B)} - E_{Bo'}^{(A-i)}} + \sum_j^{uno} \frac{(\sum_r c_r^{(j)} c_{r'}^{(o')} \gamma_{rr'}^{(o')})^2}{I_{Bo'}^{(A)} - E_{Aj}^{(B-o')}} \right\} \qquad (3.23)$$

where $0'$ denotes the singly occupied (SO) MO of B. Similarly, Eq. (3.21) becomes

$$D \sim \sum_i^{occ} \frac{c_r^{(i)\,2}}{\varepsilon_B - \varepsilon_{Ai}} \gamma_r^2 + \sum_j^{uno} \frac{c_r^{(j)\,2}}{\varepsilon_{Aj} - \varepsilon_{B'}} \gamma_r'^2 \qquad (3.24)$$

in which $\varepsilon_B = -E_{Bo'}^{(A-i)}$ and $\varepsilon_{B'} = -I_{Bo'}^{(A)}$. From the consideration of the form of γ_r and γ_r', it is worthy of note that, even in the interaction of a neutral molecule with a neutral radical, the local charge of atoms determines the magnitude of D. These equations are used for purposes which are similar to Eqs. (3.20) and (3.21).

In the case of degeneracy where one of the monotransferred configurations happens to have the same energy as the initial configuration, the first-order term of Eq. (3.10) appears. Obviously, such a case is possible only in regard to the transfer of one electron from HO MO of the donor molecule to LU MO of the acceptor molecule.

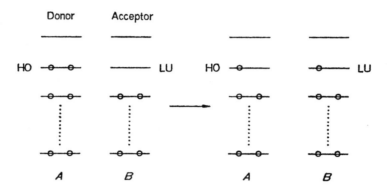

Fig. 3.3. The mode of donor-acceptor interaction

The equations corresponding to Eq. (3.14) and Eq. (3.20) are

$$D \sim |H_{0,\, \text{HO} \to \text{LU}} - S_{0,\, \text{HO} \to \text{LU}}\, H_{0,0}| \qquad (3.25\,\text{a})$$

$$D \sim \sqrt{2}\, |\sum_r c_r^{(\text{HO})}\, c_{r'}^{(\text{LU})}\, \gamma_{rr'}| \qquad (3.25\,\text{b})$$

In the special case of single-site overlapping as in Eq. (3.21), this becomes

$$D \sim \sqrt{2}\, c_r^{(\text{HO})}\, \gamma_r \qquad (\text{or } D \sim \sqrt{2}\, c_{r'}^{(\text{LU})} \gamma_{r'}) \qquad (3.26)$$

where

$$\gamma_r = c_{r'}^{(\text{LU})}\, \gamma_{rr'} \qquad (\text{or } \gamma_{r'} = c_r^{(\text{HO})}\, \gamma_{rr'}).$$

Eq. (3.25) stands for Mulliken's overlap and orientation principle. The charge-transfer interaction takes place according to the way in which the overlap of HO of the donor and LU of the acceptor becomes maximum. Particularly, the single-site interaction will occur at the position of the greatest HO density of the donor and at the position of the greatest LU density of the acceptor, as is seen from Eq. (3.26). In such cases the particular role of the frontier orbitals is evident.

Similar treatment has been made by Salem with discussions of many cases of special interest [128,129].

4. Principles Governing the Reaction Pathway

In the preceding section, the interaction energy between two reacting molecules has been discussed with the assumption of no nuclear configuration change. In the donor-acceptor interaction the delocalization stabilization is dominant. Eq. (3.25) indicates the importance of HO and LU in the donor-acceptor interaction. But the expression of Eq. (3.21) shows that in general cases the contribution of HO and LU to the quantity D is not so discriminative as those of the other MO's.

However, there exists a reason which makes the role of the frontier orbitals in the process of chemical reactions more essential than expected from the expression of D. This can be understood if the change in nuclear configuration along the reaction path is taken into consideration. The discussion of this point will be made with the aid of three principles governing the reaction pathway. A simple explanation is given in Appendix I.

i) The principle of positional parallelism between charge transfer and bond interchange

The molecular orbital has, in general, its own nodal planes. The only MO which lacks nodal planes is the lowest-energy MO; all the other MO's must have at least one nodal plane in order to be orthogonal to the lowest-energy MO.

In view of the discussion in the preceding section, the *nodal property* of HO and LU is expected to be particularly important in the theory of chemical interaction. In reality, it has already been disclosed that the nodal property of the frontier orbitals plays an essential role in determining the orientation and steric course of electrocyclic reactions [52,57,64]. Schematic diagrams for the nodal property of π HO and LU of several conjugated molecules in the frame of LCAO MO scheme are indicated in Fig. 4.1, in which shaded and unshaded areas correspond to the positive and negative regions of MO's. In the following, we can understand that this property is significant in promoting alteration of the molecular shape in case of chemical interaction.

In common molecules, an atom is as a rule bonding with neighboring atoms in each occupied MO, and antibonding in each unoccupied MO. This circumstance is seen in every example illustrated in Fig. 4.1. Also

HO   LU   HO   LU

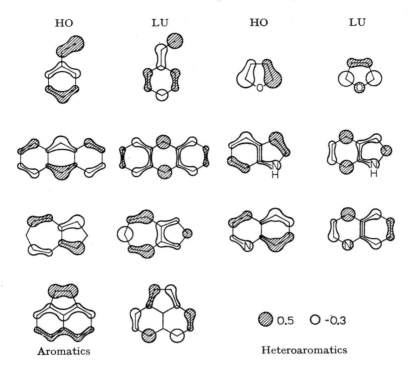

Aromatics   Heteroaromatics

⬤ 0.5   ◯ -0.3

Fig. 4.1. The nodal property of π HO and LU of some conjugated molecules

the same is easily understood by investigating the simultaneous equations which are satisfied by the LCAO coefficients and the orbital energies. For instance, if we regard the Hückel MO (all overlap integrals neglected) for the $p\pi$ electrons of planar conjugated hydrocarbons, the following relations hold with respect to the ith MO:

$$(\alpha - \varepsilon_i)\, (c_r^{(i)})^2 + \sum_{s}^{nei} c_r^{(i)}\, c_s^{(i)}\, \beta = 0 \qquad (r = 1, 2, \text{-----}) \qquad (4.1)$$

in which α is the Coulomb integral of the rth carbon $2p\pi$ AO, β is the resonance integral between neighboring $2p\pi$ AO's, ε_i is the energy of the ith MO, $c_r^{(i)}$ is the LCAO coefficient of the rth AO in the ith MO, and $\sum^{nei}$ means the summation over neighboring AO's of the rth AO. From Eq. (4.1),

$$\sum_{s}^{nei} c_r^{(i)}\, c_s^{(i)} = \frac{\alpha - \varepsilon_i}{(-\beta)}\, (c_r^{(i)})^2 \qquad (r = 1, 2, \text{-----}) \qquad (4.2)$$

23

are obtained. Since the usual hydrocarbons possess occupied MO's lower than α and unoccupied MO's higher than α, the quantity $\sum\limits_{s}^{nei} c_r^{(t)} c_s^{(t)}$ is positive for an occupied MO and negative for an unoccupied MO and is proportional to the partial "π-electron density" at the rth atom in that MO. The quantity $\sum\limits_{s}^{nei} c_r^{(t)} c_s^{(t)}$ represents the partial sum of bond orders of the rth atom with its neighbors.

Therefore, the position of the largest HO or LU density is at the same time the position where the bonds with neighboring atoms are as a whole most liable to loosening in case of electron-releasing or -accepting inter-action, respectively. Since the HO or LU density is a measure of the ease of delocalization interaction, as has been mentioned in the preceding section, this conclusion represents the parallelism between the electron delocalization and the bond interchange in a molecule in chemical reactions. Namely, the electron delocalization weakens the bonds with neighbors most at the position of the greatest frontier-orbital density.

Table 4.1. *The positional parallelism between* $(c_r^{(f)})^2$ *and* $\sum\limits_{s}^{nei} c_s^{(f)}$ *in aromatic hydrocarbons by Pariser-Parr-Pople calculation* $[(f)$ *signifies* (HO) *or* $(\text{LU})]$

Compound	position r	$(c_r^{(f)})^2$	$\pm \sum\limits_{s}^{nei} c_r^{(f)} c_s^{(f)}$ [1]
Anthracene	9	0.19472	0.08735
	1	0.09512	0.03680
	2	0.04770	0.01966
Phenanthrene	9	0.16775	0.09051
	1	0.10550	0.07154
	3	0.09868	0.06598
	4	0.05843	0.03197
	2	0.00100	0.00034
Chrysene	6	0.14446	0.07687
	1	0.07978	0.04403
	4	0.06089	0.02985
	5	0.04871	0.02809
	3	0.04871	0.02168
	2	0.01427	0.00738

[1] $+$ sign for $(f) = (\text{HO})$, and $-$ sign for $(f) = (\text{LU})$.

Qualitatively, similar relationships are ascertained in *heteroaromatic systems* where the same conclusion is derived by a numerical calculation. In more elaborate calculations than the Hückel method, such as the Pariser-Parr-Pople approximation [21,23], similar distinct parallelisms are recognized [67] (Table 4.1). Essentially the same circumstances exist also

Table 4.2. *The positional parallelism between* $(c_r^{(LU)})^2$ *and* $v_r^{(LU)}$ *of hydrogens in 2-chlorobutane by the extended Hückel calculation*

Compound	Position r	$(c_r^{(LU)})^2$	$- v_r^{(LU)}$
	5	0.07887	0.05295
	3	0.06638	0.04564
	8	0.02119	0.01700
	1	0.00978	0.00661
	7	0.00158	0.00131
	6	0.00090	0.00041
	4	0.00044	0.00011
	2	0.00029	0.00006
	9	0.00000	0.00000

$$(v_r^{(LU)} = 2 \sum_{s \neq r} c_r^{(LU)} c_s^{(LU)} s_{rs}; \; s_{rs}: \text{overlap integral})$$

in saturated compounds. This is assured [67] for instance by the extended Hückel calculation [31] (Table 4.2). Exemplifications by the various calculations mentioned above have indicated that the conclusion is independent of the level of approximation adopted, and is verified in a wide range of compounds.

ii) The principle of narrowing of inter-frontier level separation

It has been clarified that the delocalization interaction occurs at the position of the greatest frontier-orbital density which is simultaneously most susceptible to weakening of the bonds with the remaining part. This bond-weakening gives rise to a nuclear configuration change.

The direction of the nuclear configuration change is characterized by the mode of change in the energy level of HO MO of the donor and LU MO of the acceptor. The HO energy of the donor generally rises while the LU energy of the acceptor becomes lower in the event of electron delocalization, since a bonding MO is made unstable by electron-releasing while an antibonding MO is stabilized by electron-accepting, in both

25

cases through bond-weakening effectively, followed by a serious narrowing of inter-frontier energy level separation.

These circumstances become clear when we consider several common examples. The *Diels-Alder addition* of ethylene and butadiene is taken as the first and simplest example. Fig. 4.2a indicates the nodal property of HO and LU of ethylene and butadiene and the mode of delocalization interaction. The ethylene HO is bonding while LU is antibonding. The

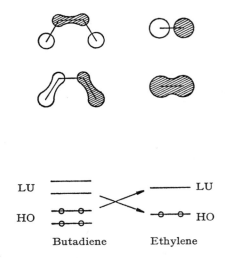

LU ——— ——— LU

HO ——— ——— HO

Butadiene   Ethylene

Fig. 4.2a. The nodal property of HO and LU in ethylene and butadiene

HO of butadiene is bonding in 1,2- and 3,4- π bonds and antibonding in the 2,3- π bonds, whereas the LU has the opposite bonding property. The electron transfer from HO of ethylene to LU of butadiene and that from HO of butadiene to LU of ethylene will both weaken the ethylene π bond and result in a double bond shift in butadiene. The change of bond lengths along the reaction path may reasonably be assumed by considering the direction of delocalization and the nodal property of frontier orbitals. It is understood in Fig. 4.2b that the changes in frontier orbital energies are remarkable, in comparison with the other MO's, so that the inter-frontier separation becomes considerably narrower as the reaction proceeds. Such relations are commonly recognized with respect to many other dienes and dienophiles [67].

Similar results are obtained also in *sigma electron systems*. Various examples can be given in regard to the S_N2 reaction of a methyl halide

26

with a halogen anion. E 2 reaction of alkyl halides, aromatic substitutions, solvation and desolvation, heterolytic addition to olefinic double bonds (see Fig. 4.3), and so on. In every reaction, the narrowing of interfrontier energy level separation between the reactant and the reagent along the reaction path is verified by numerical calculation. This implies that the importance of the frontier orbitals is more than would be expected from the case in which these circumstances are not counted.

iii) The principle of growing frontier-electron density along the reaction path

The importance of the frontier-orbital AO coefficient is evident from Eqs. (3.21) and (3.26). The problem is how this quantity changes along the reaction path. It can be shown by actual calculation that the frontier-electron density generally increases as the reaction proceeds.

A typical example is given in the case of *aromatic substitutions*. The sum of the mobile bond orders of the bonds between the reaction center

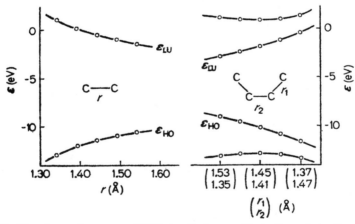

Fig. 4.2 b. The change in energy of MO's in ethylene and butadiene along the reaction path

and the neighboring atoms will gradually decrease according to the consideration stated in i) of this section, giving rise to the loosening of these bonds. The effect of this bond-loosening may be represented by a decrease in the absolute value of resonance integrals of these bonds, if the discussion is based e.g. on the Hückel MO approximation. What is to be made clear is whether or not the frontier-orbital density at the reaction

27

center would in reality increase during the process of change which is represented schematically as the following:

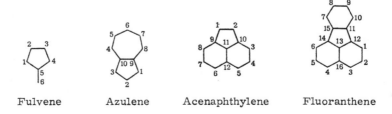

in which β is the original value of resonance integral and r stands for the position of reaction. In actual reactions the change δ is rather small (Stage II). However, in order to illustrate the general tendency of the change, an extreme case where the increment is assumptively taken as $(-\beta)$ (Stage IV) may be considered. In that case the π AO at the reaction site is ultimately isolated. In Stage III, which is reached shortly before Stage IV, a small conjugation still remains between the π AO of the reaction center and the neighboring π AO's.

In order to understand qualitatively how the frontier-electron density, $(c_r^{(HO)})^2$ and $(c_r^{(LU)})^2$, as usual grows along the path (I)$\rightarrow$(II) in planar conjugated hydrocarbons, it is convenient to take account of Stage III. In this stage it is easily proved that

$$\lim_{\Delta\beta\rightarrow 0} \{(c_r^{(HO)})^2 \quad \text{and} \quad (c_r^{(LU)})^2\} = \tfrac{1}{2} \tag{4.3}$$

provided that the hydrocarbon rest obtained by deleting the atom r from the original hydrocarbon molecule possesses one nonbonding MO, $\varepsilon = \alpha$. If the rest has n nonbonding MO's, $(c^{(HO)}_r)^2$ and $(c^{(LU)}_r)^2$ become $1/(n+1)$. Since the original frontier density values are in most cases far less than 0.5, Eq. (4.3) suggests the frontier-density growth along

Fulvene   Azulene   Acenaphthylene   Fluoranthene

the reaction path. Eq. (4.3) is valid with respect to all so-called "*alternant*" *hydrocarbons*, and also in most of the actually reactive positions of nonalternant hydrocarbons, such as $(c_1^{(HO)})^2$ of fulvene, $(c_1^{(HO)})^2$ of acenaphthylene, and $(c_7^{(HO)})^2$ of fluoranthene.

In the case in which the hydrocarbon rest has no nonbonding MO, the discussion is rather complicated [67]. In several cases it holds that

$$\lim_{\Delta\beta\to 0} (c_r^{(HO)})^2 = 1 \qquad (4.4a)$$

$$\lim_{\Delta\beta\to 0} (c_r^{(LU)})^2 = 1 \qquad (4.4b)$$

Eq. (4.4 a) is satisfied in the position 1 of azulene. Eq. (4.4 b) is valid in position 6 of fulvene, position 6 of azulene, position 3 of fluoranthene, and position 5 of acenaphthylene. Even in a few exceptional cases where the previous relations do not hold, a consideration of the coulombic effect of attacking reagents leads to a conclusion favorable to the hypothesis of frontier density growth. An example of such cases is position 3 of

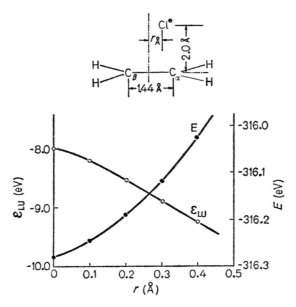

Fig. 4.3a. The change in the energy of LU, ε_{LU} and in the total energy, E, of ethylene-chlorine cation system

fluoranthene in HO MO. The rule of growing frontier density along the reaction path is essentially not violated by the adoption of more elaborate methods than the Hückel MO with respect to the calculation for aromatic substitutions.

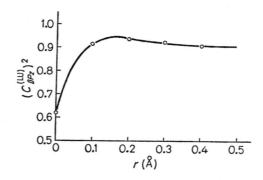

Fig. 4.3b. The changes in the LU partial population of p_z orbital at β-carbon $(c_{\beta p_z}^{(LU)})^2$, ethylene-chlorine cation system

The next example for this rule may be the *heterolytic addition of chlorine* to the C=C bond. Fig. 4.3b indicates the partial valence-inactive population [68] of the $2p_z$ AO of the β-carbon in LU, calculated by the extended Hückel method. It is seen that this quantity, $(c_{\beta p_z}^{(LU)})^2$, largely increases according to the approach of the chlorine cation to the carbon atom at which the addition is to take place, so that the reactivity of the β-position towards the second chlorine atom (anionic species) grows. Also Fig. 4.3a shows the decrease of the LU energy in the direction of the reaction path which has already been mentioned above.

5. General Orientation Rule

From the preceding discussions it is obvious that the three principles work co-operatively in promoting the reaction. As expected from Eq. (3.25), the charge-transfer interaction occurs dominantly at the position and in the direction in which the overlapping of HO and LU of the two reacting species becomes largest. The mutual electron delocalization brings about the local bond-weakening which is principally controlled by the nodal property of the frontier orbitals. The extent of the bond-loosening is positionally parallel to the frontier-orbital extension by the first principle. The weakening of bonds leads to the change in molecular shape in a definite direction, causing the narrowing of the inter-frontier energy level separation by way of the second principle, and simultaneously the frontier-electron density grows at the reaction center by the third principle. These effects will make the frontier term in the right side of Eq (3.14) more important, since the denominator becomes smaller and the numerator larger. Hence the contribution of the frontier-orbital interaction term to the delocalization part of the interaction energy (Eq. (3.14)) becomes larger, so that the amount of electron delocalization increases, again in turn resulting in promotion of bond interchange near the reaction center, molecular shape deformation, narrowing of the frontier-level separation, and frontier-density growth. In this manner, the frontier-interaction term becomes more and more significant, leading to the approximate expression

$$D \sim \frac{|H_{0,\,HO \to LU} - S_{0,\,HO \to LU} H_{0,0}|^2}{H_{HO \to LU,\,HO \to LU} - H_{0,0}} \qquad (5.1)$$

even though the charge transfer in the initial stage of interaction is not so significant as in the obvious case of donor-acceptor interaction.

It should be noted here that the MO's which can take part in such a type of co-operation are evidently restricted to the particular MO's, HO and LU. The other MO's undergo only the minimum energy change which is absolutely required for the occurrence of reaction and may reasonably be assumed to be almost constant with regard to every possible reaction site of the same sort. This is understood from the following consideration. A stable molecule originally takes the nuclear

31

configuration which is energetically most favorable. In the event of reaction, any change in nuclear configuration will bring about unstabilization. Such an unstabilization resembles the promotion in atoms in case of molecule formation. Accordingly, the change in molecular shape will occur in a direction which ensures the unstabilization is most powerfully eliminated. Any direction of change in which no energetic gain is expected will be avoided. The mutal electron delocalization between frontier orbitals gives rise to a change in molecular shape, which is thus automatically restricted to the neighbor of the reaction center in the reactant molecules. Such a *self-regulating nature* in the process of reaction will be the theoretical basis for the empirical rule which is known as "the principle of least motion" or *"the principle of least molecular deformation"* [69].

A chemical reaction is smoothly promoted by reducing the unstabilization energy ascribed to the change in molecular shape which is due to the interaction between reactant species. The most effective means of doing this is to give rise to a change by which the electron delocalization between frontier MO's is effectuated. The delocalization may be unidirectional or mutual according to the electron-donating or -accepting power of both reactants. All of the other directions of nuclear configuration change are rejected as bringing about little gain in stabilization energy.

It is thus evident that the reaction path is controlled by the frontier-orbital interaction. The position of reaction will be determined by the rule of maximum overlapping of frontier orbitals, that is, HO and LU MO's of the two reacting molecules. Sometimes SO takes the place of HO or LU in radicals or excited molecules. Hence, the general orientation principle would be as follows:

"A majority of chemical reactions are liable to take place at the position and in the direction where the overlapping of HOMO and LUMO of the respective reactants is maximum; in an electron-donating species, HOMO predominates in the overlapping interaction, whereas LUMO does so in an electron-accepting reactant; in the reacting species which have SOMO's, these play the part of HOMO or LUMO, or both."

Mention should be made here with respect to the intramolecular reactions. Some isomerization reactions, rearrangements, and the cyclization of a *conjugated olefinic chain* are the examples. The most dominant controlling factor in these cases seems to be the first-order interaction term [70,71], so that the HO—LU interaction is concealed. However, the same reaction can also be discussed by considering the frontier-orbital interaction between two parts of a molecule which are produced by a hypothetical division [72]. The HO—LU interaction has also been discussed with respect to the sigma- and pi-parts of conjugated molecules [64].

32

These two parts are regarded as if they were different molecules which are reacting with each other. A stereoselection rule which governs the reactions accompanying the hybridization change has been derived in this way. In this view, the particular MO's which seem to control the path of a chemical reaction, that is, HO,LU, and SO MO's, are referred to as *"generalized frontier orbitals"*.

The principle involved in the discussion mentioned above appears to be most general in nature, governing almost all kinds of chemical interaction, including intermolecular and intramolecular, as well as unicentric and multicentric. If the principle is applied to a unicentric reaction, it behaves as an orientation rule, and if it is employed to treat the multicentric reaction, as already mentioned in the discussion of Eq. (3.20), the stereoselection rule results [64,71,72].

It is to be noticed, however, that, considering cases like the *crystal-field* or *ligand-field interactions*, when the symmetry relationship between interacting MO's happens not to be favorable for the HO—LU interaction in a given „inflexible" configuration, the next-lying MO will temporarily act as the frontier orbital. Also in the case of *d*-orbital interaction, only the appropriate *d*-orbital which is symmetrically suitable for the interaction can play the part of the frontier orbital among the five degenerate, or almost degenerate, *d*-orbitals. The same will apply to cases of degenerate frontier orbitals (e.g. in benzene HO's and LU's) in general.

The general orientation rule described above is based solely on the consideration of delocalization interaction. Despite the discussions developed in Chap. 4, which may explain such a partiality to the delocalization term, the contribution of the other term to the interaction energy of Eq. (3.12) can never be completely disregarded. In particular, the Coulomb interaction term of Eq. (3.13) is frequently of importance. Klopman [123,124,126] took account of the effect of the first-order long-range Coulomb interaction term together with the second-order charge-transfer interaction for the purpose of discussing the chemical reactivity, introducing the concept of *"frontier-controlled"* and *"charge-controlled"* reactions. He states that to the former case belong the radical recombination and the reactions in the category of the *Woodward-Hoffmann rule* [59] as well as many conjugated hydrocarbon reactions.

6. Reactivity Indices

The reactivity index is the conventional theoretical quantity which is used as a measure of the relative rate of reactions of similar sort occurring in different positions in a molecule or in different molecules. As has already been mentioned in Chap. 2, most reactivity indices have been derived from LCAO MO calculations for unicentric reactions of planar π electron systems [73]. The theoretical indices for saturated molecules have also been put to use [58]. In the present section the discussion is limited to the indices derived from the theory developed in the preceding sections, since the other reactivity indices are presented in more detail than the frontier-electron theory in the usual textbooks [73,74] in this field.

The reactivity indices derived from the theory which has been developed in Chap. 3 are the frontier-electron density, the delocalizability, and the superdelocalizability, as has been mentioned in Chap. 2. These indices usually give predictions which are parallel with the general orientation rule mentioned in Chap. 5. The superdelocalizability is conventionally defined for the π-electron systems on the basis of Eq. (3.21) and Eq. (3.24) as a dimensionless quantity of a positive value by the following equations [57]:

i) For the reaction with an electrophilic reagent:

$$S_r^{(E)} = 2 \sum_{i}^{occ} \frac{c_r^{(i)2}}{\alpha - \varepsilon_i} (-\beta) \qquad (6.1\,\text{a})$$

Reactant   Reagent
       (Electrophile)

ii) For the reaction with a nucleophilic reagent:

$$S_r^{(N)} = 2 \sum_i^{\text{uno}} \frac{c_r^{(i)2}}{\varepsilon_i - \alpha} (-\beta) \tag{6.1b}$$

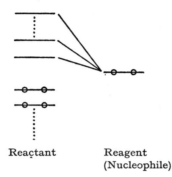

Reactant   Reagent
                (Nucleophile)

iii) For the reaction with a radical reagent:

$$S_r^{(R)} = \sum_i^{\text{occ}} \frac{c_r^{(i)2}}{\alpha - \varepsilon_i} (-\beta) + \sum_i^{\text{uno}} \frac{c_r^{(i)2}}{\varepsilon_i - \alpha} (-\beta) \tag{6.1c}$$

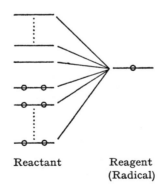

Reactant   Reagent
                (Radical)

The Hückel integrals α and β are those which have appeared in Eq. (4.1). On inspecting the form of Eq. (6.1), the conventional character involved in the definition is obvious. First of all, the problem is the orbital energy of the reagent. In order to look into the orbital energy value of a reagent, it is necessary to take account of the solvation effect in condensed-phase reactions in general. In Eq. (6.1) the reagent MO energy is taken always equal to α. Such a convention would be allowable if the previous discussions on the nuclear configuration change along the reaction pathway are taken into account. Near the transition state, the MO energy is never equal to that of the initial reagent species. Rather the charge-transfer control will cause the levelling of HO and LU MO's as the reaction proceeds:

> a too high-lying LU level of an electrophile surrounded by solvent molecules will soon descend remarkably by desolvation, and, similarly, a too low HO level of a solvated nucleophile will be elevated by the same effect.
>
> Sometimes, a too low-lying LU level of a "bare" electrophile will immediately rise by the charge transfer from reactant molecules. Similar circumstances will appear in the case of a "bare" nucleophile.

For the purpose of comparing the reactivity towards *different* reagents, however, it may be more or less recommended to take into account the effect of the reagent orbital. In that case we need to go back to Eqs. (3.22) and (3.24). Such a type of modification of superdelocalizability has also been made [58,64,75].

The introduction of hybrid-based MO's into the theoretical treatment of paraffinic hydrocarbons [58] has made it possible to extend the applicability of Eqs. (3.21) and (3.24) to a wide variety of saturated compounds. The index has been called "delocalizability" of an atomic orbital and is defined by the following formulae [58,76]:

$$D_r^{(E)} = 2 \sum_i^{\text{occ}} \frac{c_i^{(i)\,2}}{\alpha' - \varepsilon_i} (-\beta') \tag{6.2a}$$

$$D_r^{(N)} = 2 \sum_i^{\text{uno}} \frac{c_r^{(i)\,2}}{\varepsilon_i - \alpha'} (-\beta') \tag{6.2b}$$

$$D_r^{(R)} = \sum_i^{\text{occ}} \frac{c_r^{(i)\,2}}{\alpha' - \varepsilon_i} (-\beta') + \sum_i^{\text{uno}} \frac{c_r^{(i)\,2}}{\varepsilon_i - \alpha'} (-\beta') \tag{6.2c}$$

in which $c_r^{(i)}$ is the coefficient of the rth AO in the ith MO of hybrid basis, ε_i is its energy, α' is the coulomb integral of an sp^3 hybrid in a carbon atom, and β' is the resonance integral between two sp^3 hybrids in a C—C bond [76]. Sometimes the standard quantities α' and β' are referred to the sp^2 hybridized state [58]. Obviously, the two treatments are in principle equivalent.

If the effect of the reagent orbital is wanted, the term α' in Eq. (6.2) may be replaced by the reagent MO energy, α_R [58].

In the light of the discussions made in Chap. 4, the contribution of the frontier term in the formula of S_r might be more important than expected from the expression. Such a consideration has early been made and a one-term approximation of S_r (denoted by S_r') has been proposed [77]. Thus, S_r is approximated by the frontier term only:

$$S_r'^{(E)} = 2 \frac{c_r^{(HO)^2}}{\alpha - \varepsilon_{HO}} (-\beta) \tag{6.3a}$$

$$S_r'^{(N)} = 2 \frac{c_r^{(LU)^2}}{\varepsilon_{LU} - \alpha} (-\beta) \tag{6.3b}$$

$$S_r'^{(R)} = \frac{c_r^{(HO)^2}}{\alpha - \varepsilon_{HO}} (-\beta) + \frac{c_r^{(LU)^2}}{\varepsilon_{LU} - \alpha} (-\beta) \tag{6.3c}$$

Brown's reactivity index, Z-value [81], is also the one in which the frontier term solely controls the intramolecular orientation.

The contribution of the frontier orbitals would be maximized in certain special donor-acceptor reactions. The stabilization energy is represented by Eqs. (3.25) and (3.26). Even in a less extreme case, the frontier orbital contribution may be much more than in the expression of the superdelocalizability. If we adopt the approximation of Eq. (6.3), the intramolecular comparison of reactivity can be made only by the numerator value. In this way, it is understood that the frontier electron density, f_r, is qualified to be an intramolecular reactivity index. The finding of the parallelism between f_r and the experimental results has thus become the origin of the *frontier-electron theory*. The definition of f_r is hence as follows:

$$f_r^{(E)} = 2c_r^{(HO)^2} \tag{6.4a}$$

$$f_r^{(N)} = 2c_r^{(LU)^2} \tag{6.4b}$$

$$f_r^{(R)} = c_r^{(HO)^2} + c_r^{(LU)^2} \tag{6.4c}$$

In some cases half these values are adopted as f_r. The absolute value of the LCAO coefficient, $|c_r^{(i)}|$, serves as the measure of orbital extension, as well as the square value, $c_r^{(i)^2}$.

In the simple LCAO treatment in which the AO overlap is neglected, the "density" concept is rather clear-cut. An ambiguity arises in the case of inclusion of overlap. The extended Hückel calculation is one of the cases. The electron density is usually called "population" [78]. An analysis has been made with respect to the composition of population [79]. The population of the rth AO, q_r is defined by

$$q_r = 2 \sum_i^{occ} \sum_s c_r^{(i)} c_s^{(i)} s_{rs} \tag{6.5}$$

and is divided into two parts

$$q_r = p_r + v_r \tag{6.6}$$

where p_r is the "valence-inactive" part and v_r is the "valence-active" part, represented by

$$p_r = 2 \sum_i^{occ} c_r^{(i)^2} \tag{6.7}$$

$$v_r = 2 \sum_i^{occ} \sum_{s(r)} c_r^{(i)} c_s^{(i)} s_{rs} \tag{6.8}$$

in which s_{rs} is the overlap integral between the rth and sth AO's.

The value v_r is regarded as a measure of the extent to which the electron in the rth AO takes part in the bond formation with other atoms. In contrast with this, p_r is the part of population in the rth AO which is living there and responsible for the interaction with outside. Hence,

38

in view of the role of the frontier orbital in the electron delocalization interaction, it is reasonable to take, as the frontier density, the *valence-inactive part* [80]. Namely,

$$f_r^{(E)} = p_r^{(HO)} = 2\, c_r^{(HO)^2} \tag{6.9a}$$

$$f_r^{(N)} = p_r^{(LU)} = 2\, c_r^{(LU)^2} \tag{6.9b}$$

$$f_r^{(R)} = \tfrac{1}{2}\,(p_r^{(HO)} + p_r^{(LU)}) = c_r^{(HO)^2} + c_r^{(LU)^2}. \tag{6.9c}$$

Various other reactivity indices have been proposed from different theoretical bases for conjugated molecules.
The localization energy (L_r),
π electron density (q_r),
free valence (F_r),
self-polarizability (π_{rr}), and
Dewar's reactivity number (l_r) [73].
The mathematical relation between these indices and the super-delocalizability (S_r) has been disclosed [82] for alternant hydrocarbons [48a].

$$S_r = \frac{(-\beta)}{\pi} \int_{-\infty}^{+\infty} G_r(y)\,dy$$

$$L_r = \frac{1}{\pi} \int_{-\infty}^{+\infty} \log \frac{1}{y^2 G_r(y)}\,dy$$

$$F_r = \text{const.} - \frac{1}{\pi(-\beta)} \int_{-\infty}^{+\infty} \{1 - y^2 G_r(y)\}\,dy$$

$$\pi_{rr} = \frac{(-\beta)}{\pi} \int_{-\infty}^{+\infty} y^2 \{G_r(y)\}^2\,dy$$

$$l_r = \frac{2}{(-\beta)}\, G_r(0)^{-\frac{1}{2}}$$

where $G_r(y)$ is a function derived from the secular determinant. These Equations evidently reveal parallel relationships among various theoretical indices.

7. Various Examples

7.1. Qualitative Consideration of the HOMO-LUMO Interaction

The HO—LU interaction came early to the notice of theoreticians. Hückel [83] pointed out the role of LU in the alkaline reduction of naphthalene and anthracene. Moffitt [84] characterized the formation of SO_3, SO_2Cl_2, etc. by the reactions of SO_2 as an electron donor with the S-atom-localizing character of HO MO. Walsh [85] considered that the empirical result of producing nitro compounds in the reaction of the nitrite anion with the carbonium ion should be attributed to the HO of the NO_2 anion which is localized at the nitrogen atom.

Quite independently, of these fragmentary remarks, a distinctive role of HO (and later LU and SO, too) in unsaturated molecules was pointed out [51] in a general form and with substantiality (cf. Chap. 2). With respect to the molecular complex formation, the theory of charge-transfer force was proposed [55]. A clue to grasp the importance of HO—LU interaction was thus brought to light simultaneously both from the side of ionic reaction and from the side of molecular complex formation.

The Mulliken theory of overlap and orientation principle (cf. Chap. 2) [56] predicts that stabilization in the molecular complex formation should essentially be determined by the overlap of the donor HO and the acceptor LU. The *iodine complex of trimethylamine* will take the form

$$\gtrdot N -----I-I$$

since the amine HO MO is the nitrogen lone-pair orbital and the LU of iodine is an antibonding $p\sigma$ orbital extending in the direction of the molecular axis. This is also consistent with experience.

The shape of the *complex of benzene and silver cation* is also explicable in a similar manner. The HO MO's of benzene are degenerate (e_{1g}) and have the symmetry as follows:

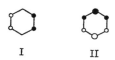

I   II

40

in which the size of circle symbolizes the orbital extension and the solid and hollow circles distinguish the different signs. Since the LU of silver cation (5s AO) is obviously spherically symmetric, the location of silver cation on the symmetry axis of benzene will nullify the HO—LU over-lapping. Hence, the cation is expected to lie above one of the C—C bonds such that

in conformity with experimental results [14]. A discussion on the silver cation complexes with various aromatic hydrocarbons has also been made [86].

A more complicated example has been discussed by Tsubomura [97]. The stability of the *quinhydrone-type complex* is ascribed to the symmetry

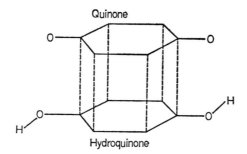

Fig. 7.1a. The quinhydrone complex

relation of HO of hydroquinone (b_1) and LU of quinone (also b_1) which is favourable for the HO—LU interaction. The orbital energy and symmetry relationship is indicated in Fig. 7.1a and b.

The same theory is useful for the understanding of the mode of orientation of ligands in many chelate compounds. The diagram [87] in

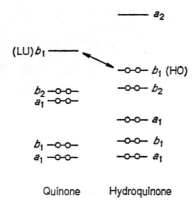

Quinone   Hydroquinone

Fig. 7.1b. The orbital symmetry relationship in quinone and hydroquinone

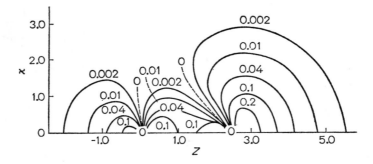

Fig. 7.2. The electron distribution diagram of HO (-0.55048 a.u.) of the CO molecule. (z: molecular axis)

Fig. 7.2 indicates the electron distribution of HO of carbon monoxide which largely localizes at the carbon atom [88]. This orbital resembles a lone-pair AO on the carbon atom and leads to the expectation that the carbon atom would behave as the electron-donating centre. As a matter of fact, the CO molecule coordinates with a metal cation by M–C–O type linkage (M represents a metal cation) in various metal carbonyl compounds. It is of interest to remark that the *total* electron population of the CO molecule has been shown by recent reliable calculation [89] to be rich on the oxygen atom in place of the carbon atom.

42

A similar result is obtained with respect to the cyanide anion CN⁻. The following mode of HO MO extension [90] underlies the M—C—N type orientation in chelate compounds:

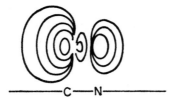

Fig. 7.3. The mode of extension of HO σ MO of CN⁻

A discussion along this line has been made in regard to the orientation of the hydrogen molecule in the *dissociative adsorption on metals* [91]. Thus, the interpretation of the function of *heterogeneous catalysis* on a molecular basis is no longer beyond our reach. The important role of LU MO in the process of polarographic reductions has also been discussed [92].

The antibonding LU MO of lithium hydride localizes more on the lithium atom than on the hydrogen atom, so that hydride anion will attack the lithium to form a linear anion.

$$H - Li + H \longrightarrow (H-Li-H)^-$$

The concept of HOMO-LUMO (or SOMO-SOMO, see Sect. 7.2) control can more generally be used for predicting the stable shape of simple molecules just like the Walsh rule [85].

Recent calculation on pyridine shows that the HO MO is not the lone-pair orbital (σ HO) but a π orbital. Nevertheless, an acceptor-like proton attacks the σ HO instead of π MO.

	93)	94)	95) 1)
π HO (a_2)	-0.44725 a.u.	$I_p(\pi) = 9.28$ eV	$a_2 > b_1 > a_1$
π NHO (b_1)	-0.45856 a.u.		
σ HO (a_1)	-0.46543 a.u.	$I_p(n) = 10.54$ eV	

NHO: next-highest occupied (orbital)
I_p: ionization potential
1) CNDO calculation

43

The reason that protonation takes place at the nitrogen lone-pair site, instead of nuclear protonation, is easily understood. In order to complete C-protonation, a large amount of energy is required for the hybridization change, whereas N-protonation does not need such energy. It is probable that a distant proton might approach the molecular plane along the extension of pi orbitals, entering then into the lone-pair region. The direction of σ protonation in pyridazine has also been discussed [96]. The result of calculation favours the configuration I.

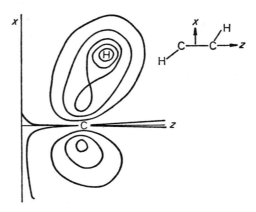

I   II

This reflects the maximum overlapping principle between HO and LU.

One of the suitable examples of sizable molecules may be *ethane*. The trans form belongs to the symmetry D_{3d}. The HO's are degenerate $1e_g$ MO, which are largely localized at C—H bonds and have bonding character on these bonds. The mode of extension is indicated below [98,99]. The LU is also localized at C—H bonds and antibonding. It is understood that most of the ionic and radical reactions of aliphatic hydrocarbons have some concern with the C—H bond.

Fig. 7.4. The HO of D_{3d} ethane. The electron distribution map in HCCH—plane

The HO of *cyclopropane* is degenerate $3e'$ MO [100]. The orbital I is responsible for a symmetric interaction, while orbital II is not. The initial interaction in protonation will take place in the σ plane as indicated. The mode of

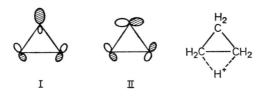

Fig. 7.5. The HO MO's of cyclopropane

conjugation of the cyclopropane ring with an adjacent π electron system[100], is of interest from theoretical point of view. The NMR study of the *cyclopropylcarbonium ion* [101] favoured the orientation (a), which is easily interpreted by the interaction of σ HO of the cyclopropyl moiety (the above-mentioned orbital II) and π LU of the dimethyl carbinyl part ($2p$ orbital).

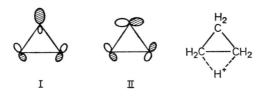

Fig. 7.6a and b. The mode of conjugation of the cyclopropane ring with the adjacent π system

Many other studies gave results consistent with similar steric configurations [102,103]. The consequence of theoretical considerations also supports the conlcusion [104,105].

The configuration of dimers of BH_3, BR_3, AlH_3, AlX_3, AlR_3, etc. may be connected to the HO and LU extensions of monomers. Literature is available with regard to the knowledge of the HO and LU of *boron hydride* [106] and *aluminum hydride* and related compounds [107].

The main cause of stability of molecules is frequently attributed to the HOMO-LUMO interaction. Several instances have been given from results of an elaborate calculation (B_2H_6,[108] BNH_6,[109] BH_3CO [110]).

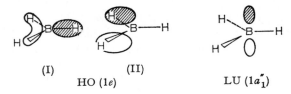

(I)   (II)

HO (1e)   LU ($1a_1'$)

Fig. 7.7. The HO and LU of boron hydride

Among many examples of *d*-orbital interaction, only the following two are selected to illustrate the feature of HO—LU conjugation. One is the *cyclooctadiene-transition metal complex*[111]. The figure indicates the symmetry-favourable mode of interaction in a nickel complex. The electron configuration of nickel is $(3d)^8 (4s)^2$. The HO and LU of nickel can be provided from the partly occupied 3*d* shell from which symmetry-allowed occupied and unoccupied *d* orbitals for interaction with cyclooctadiene orbitals are picked up.

The interaction of HO of cyclooctadiene with unoccupied *d* orbital of nickel.

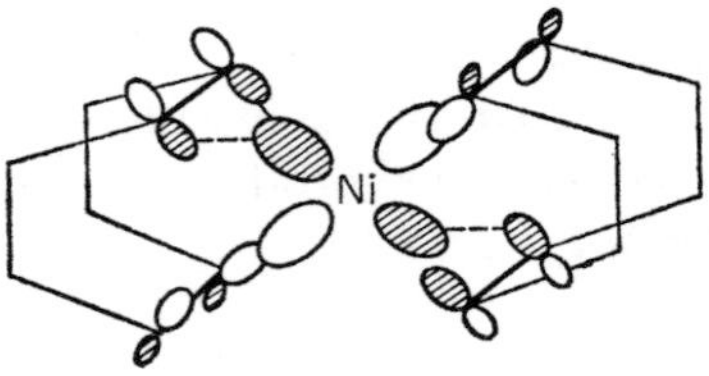

The interaction of LU of cyclooctadiene with occupied *d* orbital of nickel.

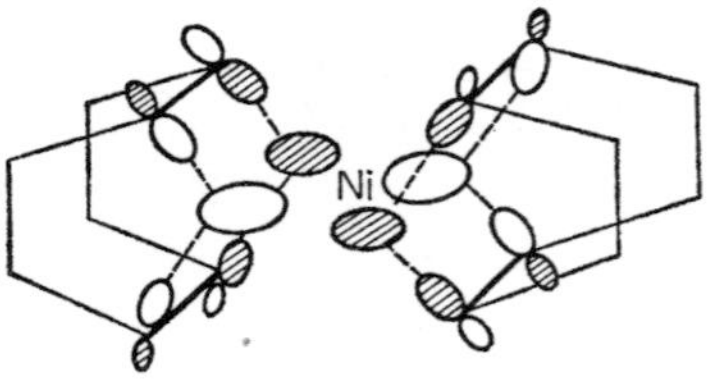

Fig. 7.8. The mode of orbital interaction in Ni-cyclooctadiene chelate

Similar chelate compounds are known, like [112]

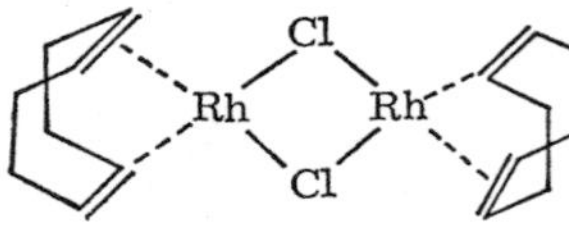

The consideration of HO—LU interaction can be used also in the interpretation of the stability of carbonium ions. For instance, the *7-norbornenyl cation* would be stabilized by the symmetry-allowed interaction

of LU of 7 methine and HO of olefinic part [113], whereas the same anion would get no such stabilization on account of symmetry prohibition (Fig. 7.9a).

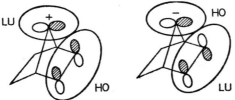

7-Norbornenyl cation   7-Norbornenyl anion

Fig. 7.9a. The conjugation stabilization in 7-norbornenyl ions

The CNDO calculation gives a result of the same trend [114].

The concept of "homoaromaticity" and "spiroconjugation" is useful for discussing the intramolecular stability of a certain unsaturated compounds. The mode of orbital interaction is indicated in Fig. 7.9b (also cf. Sect. 7.5) [113,115,116].

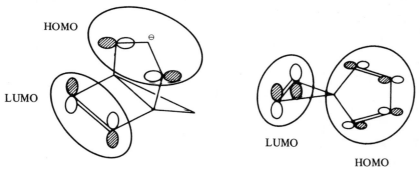

homoaromatic stabilization   spiroconjugative stabilization

Fig. 7.9b. Homoaromaticity and spiroconjugation

The favourable relationship of orbital symmetry will contribute the delocalization stabilization. Such a consideration by "partition technique" is frequently useful.

7.2. The Role of SO MO's

As has been mentioned in Chap. 5, the singly occupied MO in odd-electron molecules and radicals plays the role of HO or LU or both MO's according to the orbital energy relationship and the orbital over-

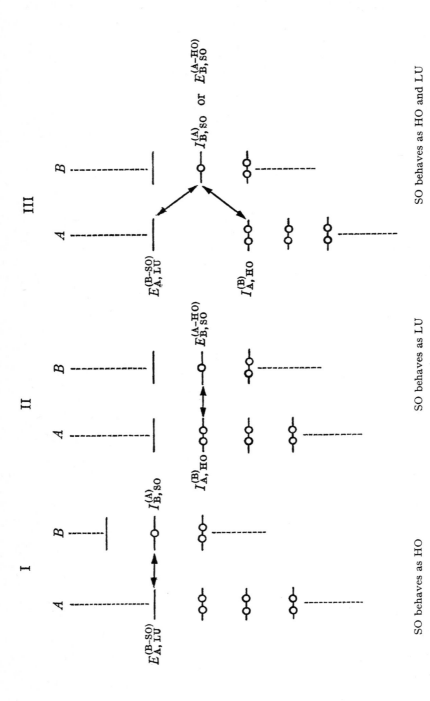

Fig. 7.10. The behavior of SO MO in interaction with a ground-state molecule[a] ($I^{(A)}_{B,SO}$, $E^{(B-SO)}_{A,LU}$, etc., denote the quantities mentioned in Chap. 3)

[a] This diagram is written in the sense of the "restricted Hartree-Fock" scheme [18]. In the "unrestricted Hartree-Fock" [19] sense each orbital of radical B is "singly" occupied and LU is higher and HO is lower than the restricted Hartree-Fock SOMO, respectively (cf. Chap. 1).

lapping situation. The importance of SO distribution is easily understood by reference to Eq. (3.23) in which the second bracket term in the right side will make a large contribution (see Fig. 7.10). Notice that even in case III in Fig. 7.10 the "mutual" charge-transfer from SO of B to LU of A and from HO of A to SO of B is of particular significance in the sense that has been mentioned in Chap. 4.

In the methyl radical, the reaction takes place in the direction of SO ($2p\pi$ of central carbon) extension, that is to say, the direction perpendicular to the molecular plane. Walsh [85] correlated the remarkable localization of SO at the nitrogen atom in NO_2 to the experimental results indicating that NO_2 abstracts hydrogen from other molecules to form HNO_2 rather than HONO, combines with NO to form $ON-NO_2$, dimerizes to produce O_2N-NO_2, and so forth. Also he pointed out that the SO MO of ClCO is highly localized at the carbon atom, which is connected with the production of Cl_2CO in the reaction with Cl_2. The SO extension of NO_2 is schematically shown below [117].

According to the recently elaborated calculation on BeH using 50-configuration wave function [118], which gives the value of -15.221 a.u. for the $^2\Sigma^+$ ground-state energy in comparison with the experimental

values of -15.254 a.u., the SO (3σ) MO of the BeH radical is largely extended in Be to the outside direction, which suggests the linear form of BeH_2 molecule (H$-$Be$-$H).

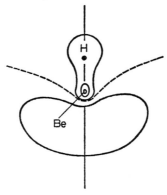

Fig. 7.11. The extension of SO MO of BeH

49

As for more complicated molecules, the exo-addition [120] in the 2-norbonyl radical was explained from the point of view of SO extension [119]. The 2-carbon is not exactly sp^2 hybridized but extends more in the exo direction than in the endo direction [125]. The nonplanarity of almost-sp^2 carbon in radicals is also expected in 2-chloroethyl and 4-t-butylcyclohexyl in which stereoselective recombinations are known. A rather exaggerated illustration of the mode of extension of SO MO is given below [121].

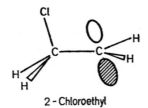

2 - Chloroethyl

Fig. 7.12. The mode of SO extension in 2-chloroethyl

A more comprehensible explanation will be given in Sec. 7.7.

The SO MO's in excited states behave in a way similar to those in radicals. Walsh [85] noticed that both of the SO MO's in the first excited state of the SO_2 molecule localize largely at the *sulphur atom*. This was correlated to the formation of bonds at the sulphur atom in the photochemical reaction of SO_2:

The theoretical significance of SO MO's in the excited-state molecules was discussed in detail [70,122]. One of these SO's, or both, play important parts in excited-state reactions.

Even in reactions involving excited states or in reactions between two radicals, the primary interaction which determines the reactivity is thought to proceed adiabatically. The probability of nonadiabatic charge transfer also may not be ignored between a molecular specie with small ionization potential and a specie with large electron affinity, in particular in the form of free, gaseous, or nonsolvated state. In that

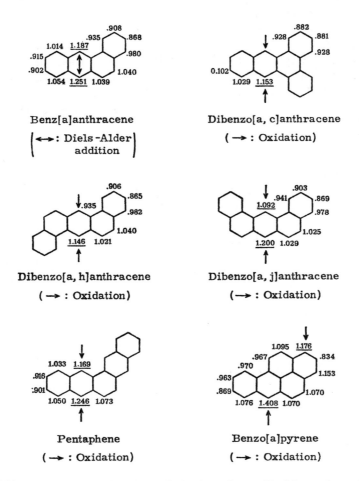

Benz[a]anthracene

$\left(\longleftrightarrow : \begin{array}{c}\text{Diels-Alder}\\ \text{addition}\end{array}\right)$

Dibenzo[a, c]anthracene

$(\longrightarrow : \text{Oxidation})$

Dibenzo[a, h]anthracene

$(\longrightarrow : \text{Oxidation})$

Dibenzo[a, j]anthracene

$(\longrightarrow : \text{Oxidation})$

Pentaphene

$(\longrightarrow : \text{Oxidation})$

Benzo[a]pyrene

$(\longrightarrow : \text{Oxidation})$

Fig. 7.13. The S_r values for aromatic hydrocarbons. (Positions of reaction are denoted by an arrow [127])

case, a zeroth order perturbation term which does not depend on the reaction position in the molecule will appear in the right side of Eq. (3.12). However, the orientation principle is not affected.

51

7.3. Aromatic Substitutions and Additions

As has already been mentioned in Chap. 2, aromatic substitution was the first object of theoretical treatment of chemical reactivity. The reactivity indices of Chap. 6 have also been first applied to the aromatic substitution. Since existing papers [51] and reviews [52,73] are available for the purpose of verifying the usefulness of the indices, f_r and S_r, only a few supplementary remarks are added here.

Fig. 7.13 is constructed from reactivity diagrams of aromatic hydrocarbons already published. The reactivity of fluoranthene has often been investigated in detail from both, the experimental and theoretical aspects [132]. The values of $f_r^{(E)}$ calculated by the Pariser-Parr method (SCF) [133] as well as by the Hückel MO (HMO) modified by considering

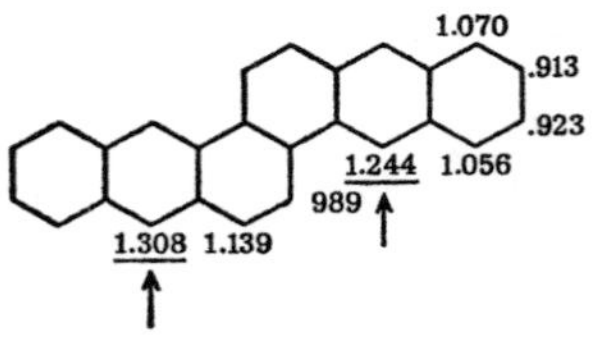

Dibenzo[b, k]chrysene

(—▸ : Oxidation)

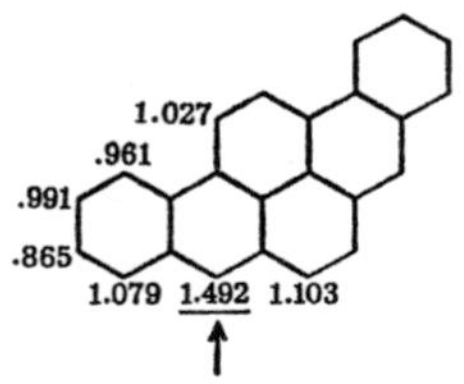

Benzo[rst]pentaphene

(—▸ : -CHO)

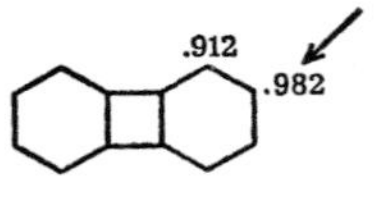

Biphenylene

(—▸ : Various reactions)

Fig. 7.13. (continued)

the next occupied MO, which lies very close to the HO, according to the appropriate procedure described in literature reference 9*b* give correctly the experimental order of reactivity $3>8>7>1>2$. The value $S_r^{(E)}$, also based on the Pariser-Parr calculation and with α put equal to the mean value of HO and LU MO energies, shows the order $3>7>8>2>1$ in slight disagreement with experiment.

These indices were initially used in the frame of Hückel MO method. But the theory has been shown to be valid also in more elaborate methods

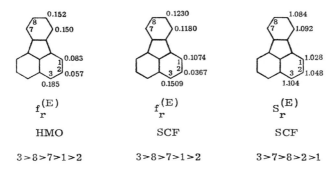

Fig. 7.14. The reactivity of fluoranthene

of calculation. Such an "approximation-invariant" character of the theory has already been discussed [52]. One of the recent examples is pyrrole. Clementi's very accurate calculation [130] gives no different result with respect to the inference of the reactive position (Fig. 7.15).

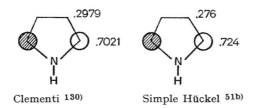

Fig. 7.15. The frontier electron density $f_r^{(E)}$ for pyrrole

A ten π electron heterocycle, *imidazo [1,2-α] pyridine* was studied by Paudler and Blewitt [131]. The protonation occurred at N_1, which was calculated to have a total π electron density less than N_4 (Fig.

7.16a). They calculated $f_r^{(E)}$ distribution to find that this is larger at N_1 than N_4 (Fig. 7.16b). Bromination took place at C_3 where both q_r and $f_r^{(E)}$ are largest.

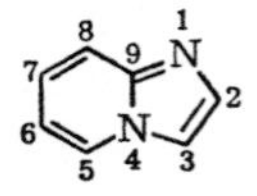

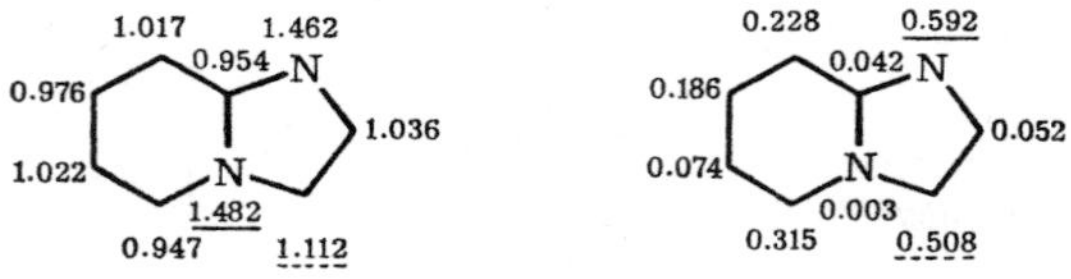

a

b

Fig. 7.16a and b. The total π electron density, q_r, and $f_r^{(E)}$ in imidazo[1,2-α]pyridine [131]. a) q_r, b) $f_r^{(E)}$

One example showing a serious "discrepancy" of the frontier electron method was reported by Dewar [134,135]. This is *10,9-borazaphenanthrene*, and the value of $f_r^{(E)}$ was reported to have been calculated by the Pople method, but the parameters used were not indicated. Fujimoto's calculation by the Pariser-Parr-Pople method [136], in perfect disagreement with Dewar's, gives the most reactive position as 8, which parallels experiment. The ambiguity involved in the integral values adopted seems to be serious, so that the establishment of parametrization for boron heterocycles is desirable.

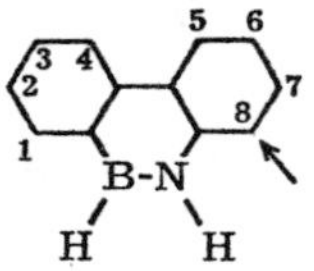

A comprehensive study has been made by the use of S_r with respect to the antioxydant action of amine compounds [137]. Several beautiful parallelisms are found between the activity and the superdelocalizability.

7.4 Reactivity of Hydrogens in Saturated Compounds

The reactivity of hydrogens at various positions of aliphatic and alicyclic hydrocarbons and their derivatives in various reactions is successfully interpreted by the theoretical indices, D_r and f_r, mentioned in Chap. 6. Most of the results obtained were reviewed in elsewhere [52,58,138] and are not repeated here.

The HO and LU MO of propane are available from the result of calculation by Katagiri and Sandorfy [29] which is based on the method already mentioned in Chap. 1. Fig. 7.17 indicates the result. Both HO and LU localize more at secondary CH bonds than at primary CH bonds, reflecting the reactivity of C_3H_8.

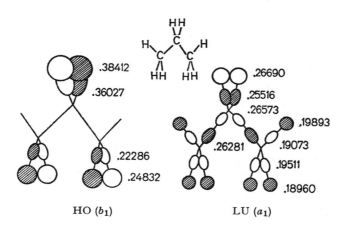

Fig. 7.17. The hybrid-based MO coefficients (absolute value) in propane. [Shaded and unshaded areas correspond to different signs of AO coefficients (+ lobe and — lobe)]

The reactivity of hydrogens in norbornane towards abstraction is of interest since the difference between two hydrogen atoms attached to the same carbon atom of position 2 can well be explained. The frontier electron density values [119] are in accord with the reactive *exo* hydrogen (Fig. 7.18).

Adamantane-type cage hydrocarbons have become a new target of theoretical investigation. The tertiary hydrogens which are known to be

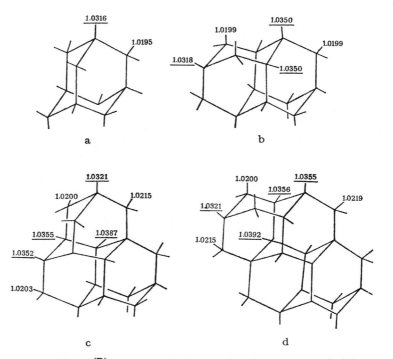

Fig. 7.18. The (HO + LU) density values of hydrogen atoms in norbornane

reactive towards homolytic are shown to have larger $D_r^{(R)}$ values than secondary ones (Fig. 7.19) [139].

Fig. 7.19 a—d. The $D_r^{(R)}$ values of hydrogens in adamantanes. a) Adamantane, b) Diamantane, c) Triamantane, d) Tetramantane

The important role of LU MO in the nucleophilic reactions of saturated hydrocarbons bearing nucleophilic substituents (halogens, alkoxy-, acyloxy-, RSO_2O-, etc.) in the molecule has been pointed out [140,141].

The LU MO of ethyl chloride (*trans* form) extends in the region of the α carbon to the direction opposite the side of the chlorine atom and also in the region of the hydrogen atom *trans* coplanar to the chlorine atom [142,149]. The former is responsible for the attack of nucleophile in S_N2 reactions, and the latter for the attack in E 2 reactions.

The value of $f_r^{(N)}$ has been calculated with respect to various *halogenoparaffins* [140,141,143]. Only one example is mentioned here. The LU density on hydrogen atoms in *t*-amyl chloride is indicated in Fig. 7.20. This MO highly localizes on *trans* hydrogens, and the hydrogen atom on C_3 has greater density than the hydrogen atom on C_1, corresponding to the reactivity of *trans* elimination and the Saytzeff rule.

Fig. 7.21 shows the example of 2-exo-chloronorbornane [141] which suggests the occurrence of the *exo-cis* elimination in conformity with experiment [144].

The S_N2 and E 2 reactions usually take place more or less concurrently.

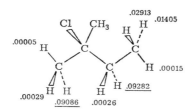

Fig. 7.20. the hydrogen $(C_r^{(LU)})^2$ values of *t*-amyl chloride

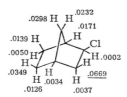

Fig. 7.21. The hydrogen $(C_r^{(LU)})^2$ values of 2-*exo*-chloronorboranane

The order of reactivity in the series of RBr is known as

$$S_N2: \quad CH_3 > C_2H_5 > (CH_3)_2CH > (CH_3)_3C$$

$$E\,2: \quad C_2H_5 < (CH_3)_2CH < (CH_3)_3C$$

which are successfully interpreted by the orbital coefficients in LU [143]. Also the base-catalyzed hydrolysis of carboxylic esters with acyl-oxygen fission can be treated in a similar fashion [143]. The LU density of protonated ketones explains the reactivity of ketones in acid-catalyzed halogenation [143].

The reaction of S_N2', that is, the *bimolecular nucleophilic substitution with allyl rearrangement*

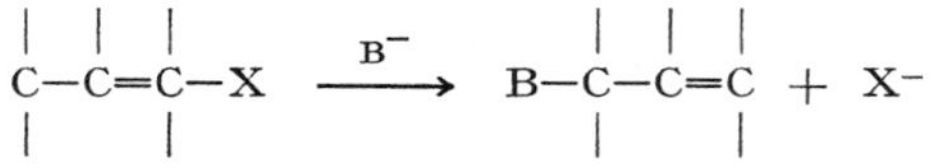

$$\underset{\substack{|\ \ |\ \ | \\ |\ \ \ \ \ |}}{C-C=C-X} \xrightarrow{B^-} \underset{\substack{|\ \ |\ \ | \\ |\ \ \ \ \ |}}{B-C-C=C} + X^-$$

is known to occur in the direction *cis* to the leaving nucleophilic group [145,146]. The LU MO of allyl chloride extends more in the direction *cis* to the chlorine atom than in the direction *trans* at the γ carbon atom [147]. The opening of the epoxy ring by the hydride anion is known to take place in the direction *trans* to the oxygen atom [148].

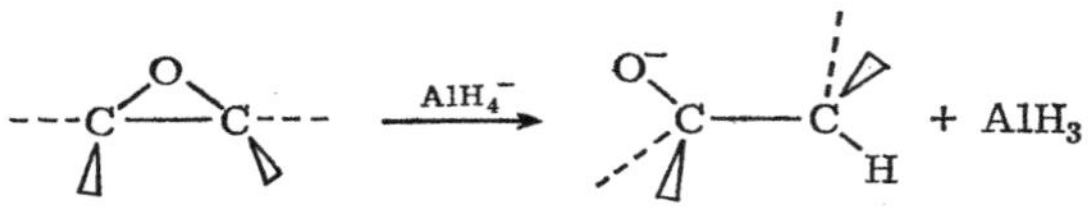

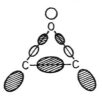

The extension of LU MO [147] explains the direction of attack of H⁻. The strong antibonding character of the C—O bond corresponds to the ring-opening reactivity.

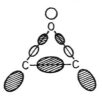

Fig. 7.22. The LU MO of ethylene oxide

The base-catalyzed allyl rearrangement of olefins can be treated by the LU orbital density criteria [150]. The LU orbital remarkably localizes at the hydrogen atoms attached to the β carbon to the double bond in various olefins, as is shown in Fig. 7.23 by the use of a few examples, and is in conformity with the experimental fact that the β hydrogen is first abstracted by the base.

The β hydrogen atoms are as a whole antibonding with the remaining part in LU MO, so that the electron delocalization to LU from the base easily comes to release these hydrogens. Similar double-bond shift reactions have also been treated [151].

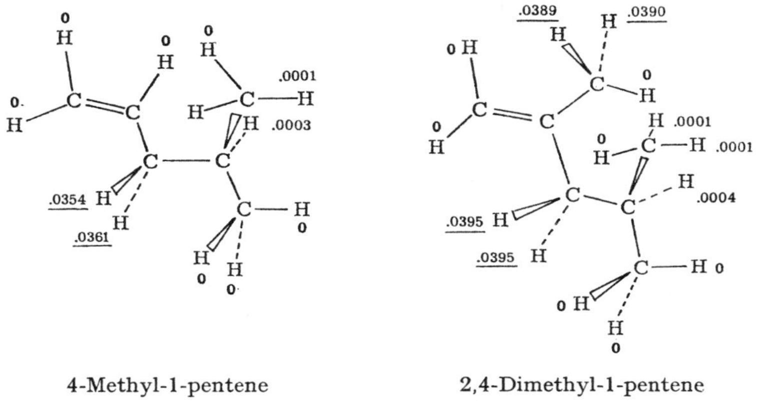
.0001
.0003
.0354
.0361
.0389
.0390
.0001
.0001
.0004
.0395
.0395
4-Methyl-1-pentene
2,4-Dimethyl-1-pentene

Fig. 7.23. The value of $2(C_r^{(LU)})^2$ in olefins

7.5 Stereoselective Reactions

In the reactions mentioned in the preceding sections, several "stereoselective" processes have been involved. Various examples have verified that the extension of singly-occupied MO determines the favorable spatial direction of interaction with other species. If there are two such nonequivalent directions in the molecule, the reaction will become stereoselective. Two or more hydrogen atoms attached to the same carbon atom are in some cases nonequivalent. Such a nonequivalence becomes a cause of stereoselectivity and has been explained theoretically. Also several cases have been mentioned in which some nucleophiles selectively attack the molecule from a certain spatial direction.

The general relation which must be satisfied in order to bring about an appreciable stabilization energy in the chemical interaction has been given by Eq. (3.20) and Eq. (3.25b). Such relations frequently provide a "selection rule" for the occurrence of stereoselective reactions.

Such a selection rule was first found in the *Diels-Alder addition* [52]. Eq. (3.25 b) is simply applied to the interaction between the HO of dienes and the LU of dienophiles, obtaining

$$D \sim \sqrt{2} \mid c_r^{(HO)} \, c_{r'}^{(LU)} + c_s^{(HO)} \, c_{s'}^{(LU)} \mid \cdot \mid \gamma \mid \qquad (7.1)$$

where $\gamma_{rr'}$ and $\gamma_{ss'}$ are taken to be equal, and r and s denote the 1,4-positions of the diene, and r' and s' the corresponding 1,2-positions of the dienophile (Fig. 7.24).

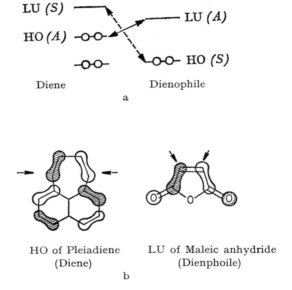

Fig. 7.24 a and b. The diene-dienophile interaction. a) Orbital relationship, b) An example. (S: symmetric, A: antisymmetric)

It is easily understood from Eq. (7.1) that the signs of $\{c_r^{(HO)} \, c_{r'}^{(LU)}\}$ and $\{c_s^{(HO)} \, c_{s'}^{(LU)}\}$ are required to be the same in order for D to have an appreciable magnitude. All the examples of combination of diene and dienophile in which reaction actually takes place were found to satisfy this condition [52]. It is to be noticed that this conclusion is independent on the sign of each AO adopted (Return to Eq. (3.20)).

Similar relationships have been established with regard to the *1,3-dipolar addition* and the *photodimerization of olefins* [70]. The HO—LU

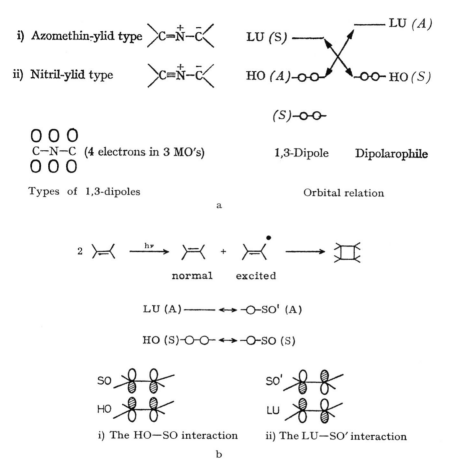

Fig. 7.25a and b. Orbital relationships in 1,3-dipolar additions and photodimerization of olefins. a) 1,3-Dipolar addition, b) Photodimerization of olefins

symmetry relationship is favourable for the overlap stabilization, as is seen in Fig. 7.25. The essential part of 1,3-dipoles is a four-π electron system with three π AO's, and the orbital symmetry relation is in favour of the interaction of 1 and 3 positions of 1,3-dipole with 1 and 2 positions of dipolarophile, respectively. The photodimerization of olefins, in which one reactant is thought to be an excited molecule, may proceed by way

61

of a similar favourable orbital symmetry relationship. Other cyclic dimerizations and cyclic rearrangements like Claisen and Cope types were similarly treated [70].

Theoretical considerations in the same fashion enable predication of the possible configuration of the transition state. Eq. (3.25b) for the multicentre interaction is utilized. Hoffmann and Woodward [152] used such methods to explain the endo-exo selectivity of the Diels-Alder reaction (Fig. 7.26). The maximum overlapping criteria of the Alder rule is in this case valid. The prevalence of the endo-addition is experimentally known [153].

Similar discussion is possible with respect to the transition state of the *Claisen* and *Cope rearrangements* [154]. These can be treated similarly. Fig. 7.27a indicates that the symmetry of SO MO suggests *cis-cis* interaction with the six-membered structure for the transition state, but the chair-boat selectivity is not determined by the SO—SO interaction. The overlapping of LU' and HO' plays a secondary role. Fig. 7.27 shows that the boat form is unfavourable in comparison with chair form on account of the different signs of LU' and HO' at the central carbons. Similar consideration is possible with respect to the extended Cope rearrangement (Fig. 7.27 b). The predominance of the chair-form transition state is known both in the Claisen [155] and the Cope rearrangements [156].

The theory of stereoselectivity found in intramolecular hydrogen migration in *polyenes* was disclosed by Woodward and Hoffmann [59]. The HO—LU interaction criterion is very conveniently applied to this problem [72]. The LU of the C—H sigma part participates in interaction with the HO of the polyene π part. The mode of explanation is clear-cut

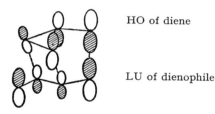

HO of diene

LU of dienophile

Endo-configuration

Fig. 7.26. The exo-endo selectivity in Diels-Alder reactions

in Fig. 7.28. For instance, in case a), when the carbon hybrid at the C—H σ part is given the same sign as the end π AO of the butadiene part, the

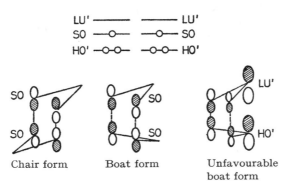

Chair form   Boat form   Unfavourable
boat form

Fig. 7.27a. The transition state in the Cope rearrangements. Cope rearrangement

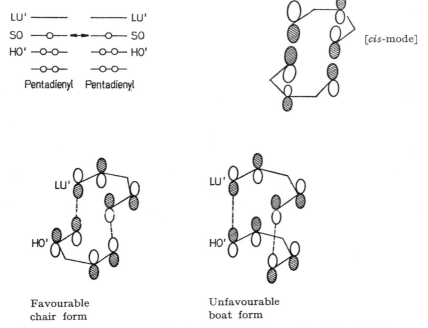

Favourable   Unfavourable
chair form   boat form

Fig. 7.27b. The transition state in the Cope rearrangements. Extended Cope rearrangement

sign of the hydrogen atom comes to have the same sign as the *upper* half (the *same* side part as the H atom with respect to the molecular plane) of the π AO of the other end of butadiene. The "selection rule" derived from Eq. (7.1) is thus satisfied so that the hydrogen can migrate "supra-

63

facially". The thermal 1,5-migrations are experimentally known. On the contrary, thermal 1,7-transfer does not fulfil this requirement. Except in very special cases [157], such "antarafacial" displacements are not actually known. The relation is reversed in the case of photochemical processes which are considered to occur in the lowest excited state of the polyene π part. The 1,7-migration is favourable in this case. Many experimental evidences are mentioned in reference [59].

a) Ground-State Reaction

i) 1,5-migration
   (in general
   $1,(4n+1)$-transfer)

ii) 1,7-migration
   (in general
   $1,(4n+3)$-transfer)

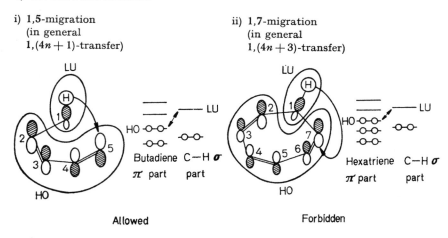

b) Excited-State Reaction

i) 1,7-migration
   $(1,4n+1)$-transfer)

ii) 1,5-migration
   $(1,(4n+3)$-transfer)

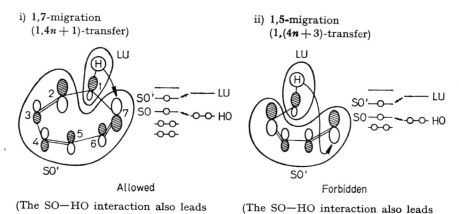

(The SO—HO interaction also leads
to the same conclusion)

(The SO—HO interaction also leads
to the same conclusion)

Fig. 7.28. The selection rule for the hydrogen migration in olefins

The cyclization of conjugated polyenes and the inverse reaction were those processes which provided superb materials [158] leading to the Woodward-Hoffmann rule [59].

The energy change, ΔE, due to a *new "bond"* arising between two $2p\pi$ AO's of a conjugated hydrocarbon, r and s, is simply represented by the Hückel calculation as

$$\Delta E = 2\,P_{rs}\,\gamma \qquad\qquad (7.2)$$

in which P_{rs} is the π bond order between rth and sth π AO's and γ is the "resonance" integral between these two AO's. Of course, the sign of γ depends on the sign of π AO's adopted. If the signs of π AO's are taken, as used to be, as indicated in Fig. 7.29a, where each $2\,p\pi$ AO has the same sign in the same side of molecular plane, there may be two possible cases of interaction between two π AO's, which are illustrated in Fig. 7.29b. In Type I interaction the two π AO's overlap in the region of the same sign, whereas Type II overlapping is concerned with the regions of the different signs. In the former case the resonance integral is negative, while in the latter case it becomes positive. Eq. (7.2) shows that the

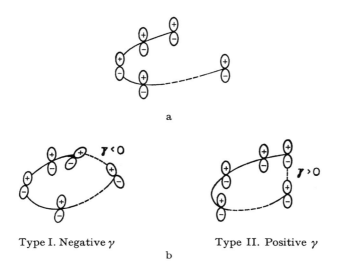

a

Type I. Negative γ Type II. Positive γ

b

Fig. 7.29a and b. The relation between the mode of cyclization and the sign of π AO's. a) The assignment of signs to AO's in conjugated polyenes, b) Two modes of σ-type interaction between two AO's

stabilization due to this overlapping occurs in the case of $P_{rs} > 0$, when $\gamma < 0$, and in the case of $P_{rs} < 0$, when $\gamma > 0$, since the perturbation in the Hamiltonian operator is considered to be negative. Therefore, the selection of the two modes of interaction depends on the sign of π bond order between the two π AO's.

The sign of P_{rs} with respect to various pairs of positions in conjugated polyenes in their ground state was investigated [159]. The result is indicated in Fig. 7.30. The π bond order is always positive in the cases where

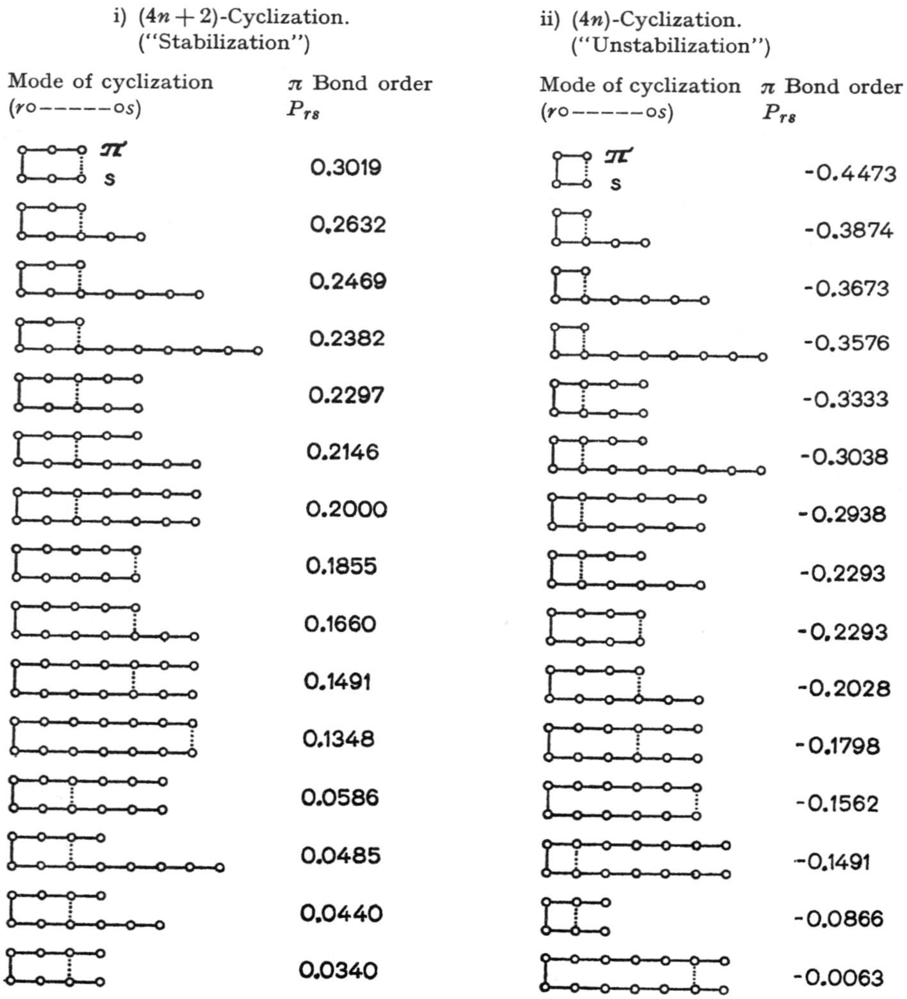

i) $(4n+2)$-Cyclization. ("Stabilization")		ii) $(4n)$-Cyclization. ("Unstabilization")	
Mode of cyclization $(ro\,\text{-----}\,os)$	π Bond order P_{rs}	Mode of cyclization $(ro\,\text{-----}\,os)$	π Bond order P_{rs}
	0.3019		-0.4473
	0.2632		-0.3874
	0.2469		-0.3673
	0.2382		-0.3576
	0.2297		-0.3333
	0.2146		-0.3038
	0.2000		-0.2938
	0.1855		-0.2293
	0.1660		-0.2293
	0.1491		-0.2028
	0.1348		-0.1798
	0.0586		-0.1562
	0.0485		-0.1491
	0.0440		-0.0866
	0.0340		-0.0063

Fig. 7.30. The cyclization of conjugated polyenes [159]

$r-s$ cyclization might form a $(4n+2)$-cycle, while it is always negative in the formation of $(4n)$-cycle. Hence, $(4n+2)$-cyclization would take place by way of Type I interaction, and $(4n)$-cyclization through Type II interaction [70,71]. Mathematical formulations were made for such explanations [70,160].

Woodward and Hoffmann have first disclosed that the thermal $(4n+2)$-cyclization (and also the photochemical $(4n)$-cyclization) takes place via Type I process, and the photochemical $(4n+2)$-cyclization (and also the thermal $(4n)$-cyclization) via Type II process [59]. They called the former (Type I) process "*disrotatory*", while the latter (Type II) process was referred to as "*conrotatory*". They attributed this difference in selectivity to the symmetry of HO and SO' MO in the ground-state and excited-state polyene molecules, respectively (Fig. 7.31). The former is symmetric with respect to the middle of the chain, and the latter antisymmetric, so that the intramolecular overlapping of the end regions having the *same* sign might lead to the Type I and Type II interactions, respectively.

The reverse process to cyclization, that is, the *ring-opening* of cyclic polyenes was discussed simultaneously by Woodward and Hoffmann [59]. Here we might adopt another way of reasoning which is consistent with the discussions made since the beginning of this section. The mechanism of ring cleavage is understood by considering the participation of the

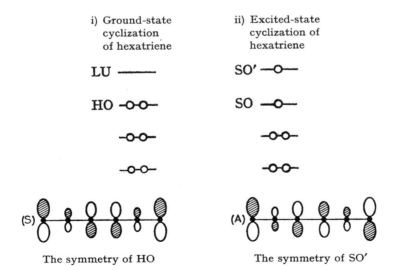

i) Ground-state cyclization of hexatriene

ii) Excited-state cyclization of hexatriene

The symmetry of HO

The symmetry of SO'

Fig. 7.31. The symmetry of the highest occupied MO in the groundstate and the excited-state hexatriene molecules

C—C σ bond to be cleft (Fig. 7.32). The LU of the C—C σ part will conjugate with the HO of the π part of the ground-state polyene moiety in case of reaction, so that the orbital symmetry relations clearly determines the direction of bond fission. The direction of change is indicated by arrows. In this manner, in the thermal opening, the ($4n$)-chain will be

i) Thermal opening

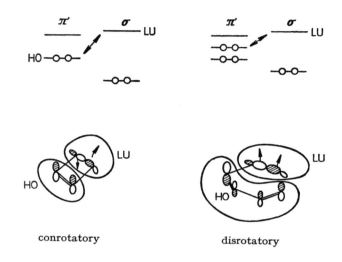

conrotatory   disrotatory

ii) Photochemical opening

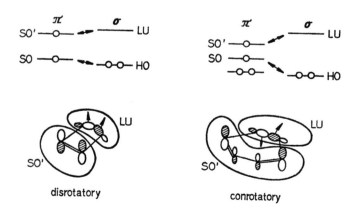

disrotatory   conrotatory

Fig. 7.32. The steric pathway of the ring-opening of cyclic polyenes. (In case ii) the consideration of the SO—HO interaction does not change the conclusion)

formed by a conrotatory pathway, whereas the $(4n+2)$-chain will be produced through a disrotatory process. In the photochemical cleavage, on the contrary, the $(4n)$-chain formation will proceed by a disrotatory fashion and the $(4n+2)$-chain formation by a conrotatory mode. Such a conclusion is most easily derived by the relation of Eq. (7.1), if we investigate the direction of arrows indicated in Fig. 7.32.

It is of interest to investigate the usefulness of this theory to the chemical change involving the interaction between the σ and π parts of conjugated systems [64,70,161]. Such σ-π interactions are frequently stereoselective. The addition to olefinic double bonds and the α,β-elimination are liable to take place with the *trans*-mode [162]. The Diels-Alder reaction occurs with the *cis*-fashion with respect to both diene and dienophile.

The mode of σ-π interaction is classified into *syn*- and *anti*-interactions. These are defined as indicated in Fig. 7.33. The carbon atoms initially sp^2-hybridized change into the sp^x-hybridized state where x is a number between 2 and 3.

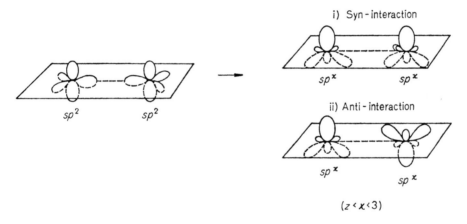

Fig. 7.33. The two modes of a two-centre σ—π interaction

The α,β-*noncycloaddition to an ethylenic bond* and α,β-*elimination* are taken as the first example. The σ and π parts are regarded as if they were two separate molecules. The direction of change in hybridization is dominated by the overlapping of LU of σ part (Fig. 7.34a) and HO of π part (Fig. 7.34b), so that the mode of interaction becomes as indicated in Fig. 7.34c.

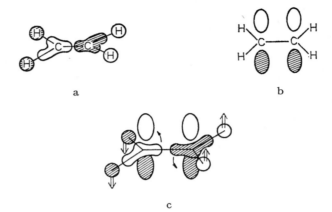

a   b

c

Fig. 7.34a—c. The α,β-noncyclic σ—π interaction. a) σ LU MO, b) π HO MO,
c) The direction of interaction [Anti-mode]

→ : The direction of σ AO mixing
⇒ : The direction of nuclear configuration change

Similar treatment explains the prevalence of the *syn*-mode [163,164] in
α,γ-interaction in S_N2' reactions (Fig. 7.35a). The α,δ-interaction (e.g.
S_N2' type reaction) is predicted to occur with *syn*-mode, and α,ε-interaction with *anti*-mode (Fig. 7.35b and c).

Also the σ-π interaction in Diels-Alder additions, which occurs with
syn-fashion with regard to both diene and dienophile, is explained (Fig.
7.36). For the first place, the p-σ type interaction is allowed, by the selection rule already mentioned, between the π-part of butadiene and the
π-part of ethylene. Once this weak p-σ type interaction starts, the p AO
part forms a six-electron system. The HO of this p-part will come from
HO of butadiene π-part interacting with LU of ethylene π-part will
interact with σ-LU's of both butadiene and ethylene. The mode of interaction is as indicated in Fig. 7.36.

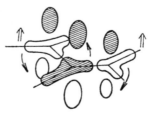

π HO and σ LU
of allyl cation

a

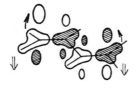

π HO and σ LU
of butadiene

b

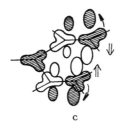

π HO and σ LU
of pentadienyl cation

c

Fig. 7.35a—c. The α,γ-, $\alpha\delta$-, and $\alpha,\varepsilon - \sigma,\pi$-interactions

a) α,γ-Interaction [*Syn*-model]
b) α,δ-Interaction [*Syn*-model]
c) α,ε-Interaction [*Anti*-model]

$\rightarrow$: The direction of σ AO mixing
$\Rightarrow$: The direction of nuclear configuration change
The hydrogen AO's are not indicated here

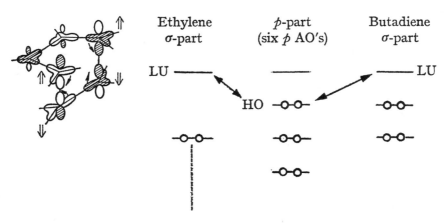

Fig. 7.36. The mode of $\sigma-\pi$ interaction in Diels-Alder reactions

$\rightarrow$: The direction of $\sigma-$AO mixing
$\Rightarrow$: The direction of nuclear configuration change
The hydrogen AO's are not indicated here

71

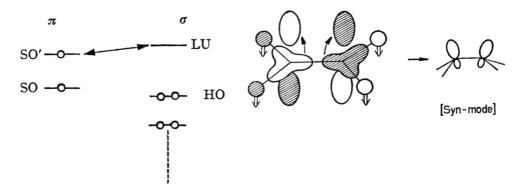

Fig. 7.37. The 1,2-addition to the excited-state ethylenic bond

The σ-π interaction in the excited-state π electron systems is also successfully treated. The 1,2-addition will take place with *cis* mode as is indicated in Fig. 7.37. This was predicated in reference [64]. Experimental evidence [72,165] is the photoinduced addition of *N*-chlorourethane to olefins which gives mainly *cis* addition product, while thermal addition produces a dominantly *trans* adduct.

An interesting example of the application of the theory is a prediction of a new route to *polyamantane* by polymerization of *p*-quinodimethane [139]. The first step would be π-π overlapping interaction. The HO and LU of quinodimethane are indicated in Fig. 7.38a. The mode of π HO-LU interaction and the possible structure of polyamantane derived therefrom (Type I polymer) can be seen in Fig. 7.38b. On the other hand, the direction of the hybridization change would be controlled by the σ-π interaction. The nodal property of π HO and σ LU of the monomeric unit are as shown in Fig. 7.38c, so that the hybridized states of carbon atoms might change into the form illustrated in Fig. 7.38d to lead to the Type II polymer.

Miscellaneous examples of σ-π interactions are listed in the following and in Fig. 7.39. The theoretical conclusion serves in some cases as the explanation of experience in relation to the direction of stereoselection and in some cases as prediction.

a) The HO and LU in the pi-electronic part of *p*-quinodimethane

HO

LU

HO

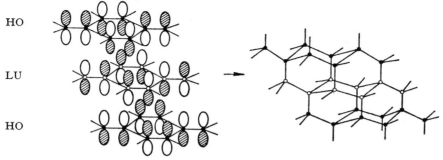

[Type I polymer]

b) The mode of π HO—LU interaction and the possible structure of polyamantane

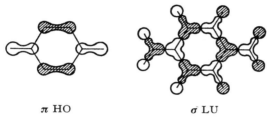

π HO   σ LU

c) The nodal property of π HO and σ LU

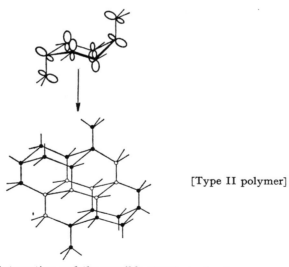

[Type II polymer]

d) The mode of σ—π interaction and the possible structure of polymer

Fig. 7.38a—d. A prediction of a new route to polyamantane on the basis of orbital symmetry consideration. (The shaded and unshaded areas correspond to the positive and negative regions of the wave functions, respectively)

Various Examples

a) Cope and Claisen rearrangements

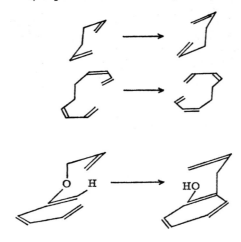

b) Bicycloheptene rearrangement [166)]

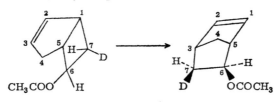

c) Deamination of cyclic imines [167)]

These reactions seem to take place through participation of the σ bond to be split in conjugation with the π part. The direction of the bond fission is indicated in Fig. 7.39 by arrows.

The nitrogen extrusion reaction of various cyclic azo compounds has been classified and explained by the use of HOMO—LUMO interaction scheme [173].

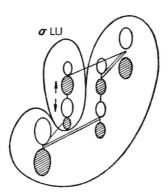

HO of a system consisting of two ethylenes connected by a weak $p\sigma$ bond

a) (i) Cope rearrangement

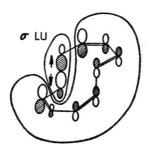

HO of a system consisting of two butadienes connected by a weak $p\sigma$ bond

a) (ii) Extended Cope rearrangement

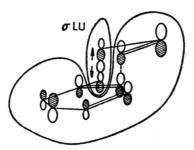

HO of a system consisting of ethylene and benzene connected by a weak $p\sigma$ bond

a) (iii) Claisen rearrangement

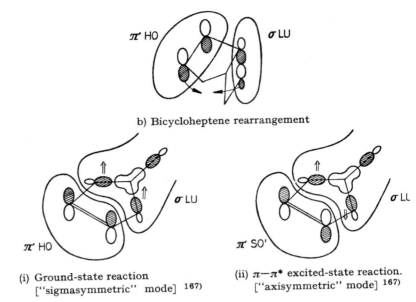

b) Bicycloheptene rearrangement

(i) Ground-state reaction
["sigmasymmetric" mode] [167]

(ii) π—π* excited-state reaction.
["axisymmetric" mode] [167]

c) Deamination of cyclic imines

Fig. 7.39a—c. Various examples of σ—π interaction. (Arrows indicate the direction of σ bond fission)

7.6. Subsidiary Effect

Fig. 7.40 shows the influence of Lewis acid catalysts on the endo-exo selectivity in the Diels-Alder reaction of acrolein and dienes [174]. The LUMO amplitude at the carbonyl carbon grows by the influence of the protonated carbonyl oxygen, increasing the subsidiary effect which favors the endo pathway.

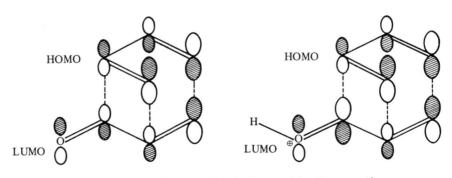

Fig. 7.40. Effect of protonation in the acrolein-diene reaction

The orbital phase relation favors the endo path in the thermal dimerization of cyclobutadiene also, as is indicated in Fig. 7.41 [175].

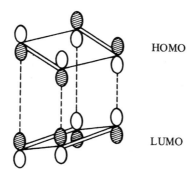

Fig. 7.41. Thermal dimerization of cyclobutadiene

In Diels-Alder reactions the orientation is affected by the substituents. The reactivity of the dienes with electron-attracting substituents is governed by the interaction between diene-HOMO and dienophile-LUMO. It is assumed that the orientation is determined by the overlapping of HOMO and LUMO at maximum lobes [176]. Fig. 7.42 shows an example of such *regioselectivity*.

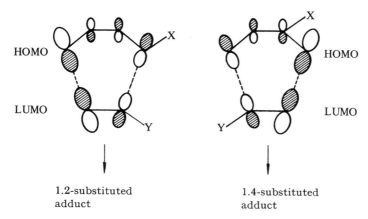

Fig. 7.42. Regioselectivity in Diels-Alder reactions of substituted dienes and dienophiles
(X: electron-repelling substituent)
(Y: electron-attracting substituent)

Similar regioselectivity in 1.3-dipolar additions is exemplified in Fig. 7.43 [177].

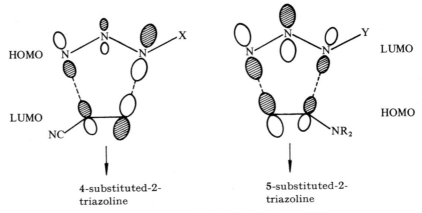

HOMO

LUMO

NC

LUMO

HOMO

NR$_2$

4-substituted-2-
triazoline

5-substituted-2-
triazoline

Fig. 7.43. Regioselectivity in 1.3-dipolar additions
(X: electron-repelling substituent)
(Y: electron-attracting substituent)

Houk's proposition mentioned above [174,176,177] is supported by a theoretical calculation with regard to the reaction of ethylene and diazomethane [178].

The reaction of 1.3-dipole

$$a \cdots b^+ \longrightarrow c^-$$

with fulvene gives two different types of product, I and II.

I

I—1

I—2

II

II—1

II—2

In general, ylid-type 1.3-dipoles afford type I products while imine- or oxide-type dipoles give type II compounds. The selectivity between these two types is referred to as *"periselectivity"*[179]. The selectivity between I-1 and I-2, or between II-1 and II-2, is the already mentioned regioselectivity. The periselectivity is excellently interpreted by Houk with the use of HOMO-LUMO interaction criterion. Ylid-type 1.3-dipoles have high HOMO and LUMO levels and are called "HOMO-controlled", while imines and oxides have low HOMO and LUMO levels and are named as "LUMO-controlled". The orbital phase of fulvene HOMO and LUMO leads to the results shown in Fig. 7.44.

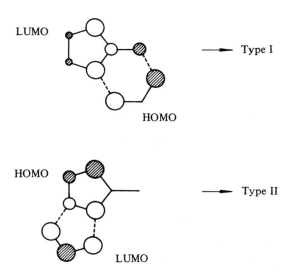

Fig. 7.44. Periselectivity in the reaction of fulvene and 1,3-dipoles

7.7. Rehybridization by Neighboring Group Effect

Let two orbitals of a molecule A be ψ_{Ai} and ψ_{Aj}. These may be concentric carbon 2s and 2p orbitals, or σ and π MO's of a planar molecule.

An example is the explanation of the experimentally wellknown *exo*-selectivity of 2-norbornyl radical [168]. We consider the influence of the sigma-bonding orbital of methylene bridge upon the odd electron orbital by the criterion mentioned in Appendix II, (AII-5), case ii) (Fig. 7.45a).

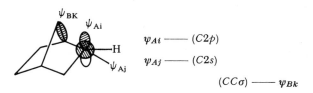

$$\psi_{Ai} \text{——— } (C2p)$$
$$\psi_{Aj} \text{——— } (C2s)$$
$$(CC\sigma) \text{——— } \psi_{Bk}$$

Fig. 7.45a. Orbital interaction in norbornyl

The resulting modified $C2p$ orbital is $[\psi_{Ai} + \psi_{Aj} - \psi_{Bk}]$ corresponding to a partial extension to exo direction (Fig. 7.45 b).

Fig. 7.45 b. The extension of odd-electron orbital in norbornyl

The preference of exo-reactivity in norbornene and norbornenone [169] seems to belong to the same category.

Another similar instance is the exo-selectivity in bicyclopentene (Fig. 7.46 a) [170]. The influencing orbital is the high-energy bent bond of cyclopropane ring. A similar selectivity is observed in bicyclo [3.1.0] hexene (Fig. 7.46 b) [171].

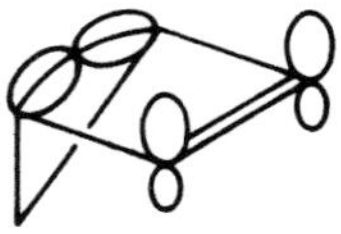

Fig. 7.46a 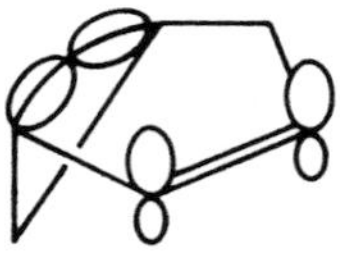   Fig. 7.46b

In the stable chair conformation of cyclohexanone having a large group at 4 position the nucleophilic addition to carbonyl takes place in the axial direction preferentially (Fig. 7.46 c) [172].

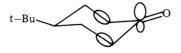

Fig. 7.46c

Fig. 7.46. Nonequivalent orbital extension in cyclic unsaturated compounds

8. Singlet-Triplet Selectivity

The reactions of excited states usually occur in the singlet state and in the triplet state. Consider an open-shell electron configuration in which two orbitals a and b have one electron each.

The singlet- and triplet-state energies are written as

$$^1W = \{h_{aa} + h_{bb} + 2sh_{ab} + (aa|bb) + (ab|ab)\} / (1 + s^2)$$
$$^3W = \{h_{aa} + h_{bb} - 2sh_{ab} + (aa|bb) - (ab|ab)\} / (1 - s^2)$$

$$(8.1)$$

where $s = \int a(1)b(1)\,dv(1)$, $h_{ab} = \int a(1)h(1)b(1)\,dv(1)$, $(ab|cd) = \iint a(1) b(1) \dfrac{1}{r_{12}} c(2)d(2)\,dv(1)\,dv(2)$, and h is the one-electron Hamiltonian. If these two orbitals localize in separate regions with a small overlap integral, it follows that

$$^1W \sim h_{aa} + h_{bb} + (aa|bb) + s^2 \text{ (negative quantity)}$$
$$^3W \sim h_{aa} + h_{bb} + (aa|bb) - s^2 \text{ (negative quantity)}$$

$$(8.2)$$

Therefore, a singlet state stabilizes with more overlapping, a triplet state with less. Accordingly, two odd electrons tend to approach each other in a singlet state whereas they are apt to go away from each other in a triplet state resulting in a characteristic feature of the triplet state reaction [180].

The photoinduced reaction of 1.3-pentadiene [181] produces 1-methyl-cyclobutene-2, 1.4-pentadiene, and 1.3-dimethylcyclopropene in the singlet-state reaction while in the triplet state only 1.3-dimethylcyclo-propene is formed. The orbital interaction scheme can easily interpret this result, as is shown in Fig. 8.1. In the triplet state a cyclic biradical

81

is formed so that among two odd electrons one is situated on the ring while the other is out of the ring. In the singlet state the two odd electrons are free to come near.

i) Singlet-state reaction

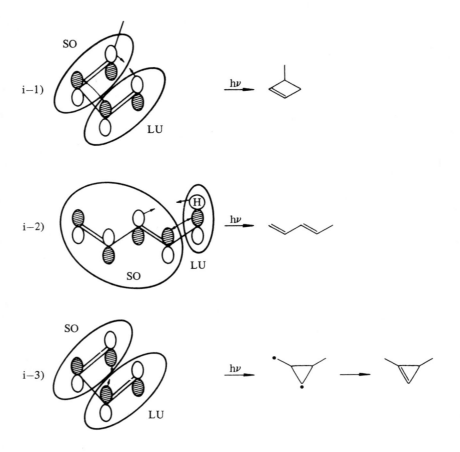

ii) Triplet-state reaction (same as i-3)

Fig. 8.1. Photoinduced reaction of 1.3-pentadiene

Another example is the photochemical reaction of β-, γ-unsaturated ketones (Fig. 8.2) [182].

i) Singlet-state reaction

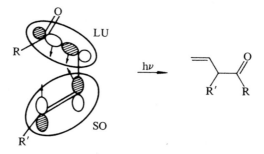

ii) Triplet-state reaction

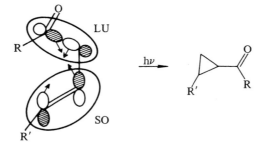

Fig. 8.2. Photoinduced reaction of β-, γ-unsaturated ketones

9. Pseudoexcitation [183)]

The orbital phase relation of HO and LU of monoolefins does not favor the concerted thermal dimerization, while the photoexcitation facilitates the olefin dimerization to cyclobutanes, as has been touched on in Sect. 7.5. An unsaturated bond under the influence of a strong electron acceptor, however, frequently adds easily to an olefin to form a four-membered ring [184)]. For instance

($^1\Delta_g$ Oxygen)

(Benzyne)

The reactions of olefins with carbon dioxide, formaldehyde, tetra-cyanoethylene, azodicarboxylic ester, ketene, keteneimmonium ion, chlorosulfonyl isocyanate, and so on, belong to the same category.

A theoretical study suggested the appropriateness of the three-center interaction model (I) for the initial stage (Fig. 9.1) [185)]. This model is reasonable also from the HO—LU orbital-interaction scheme (II) in view of the donor-acceptor relationship.

In the subsequent stage the four-membered ring is brought to being steered by the quasi-four-center orbital interaction (III).

84

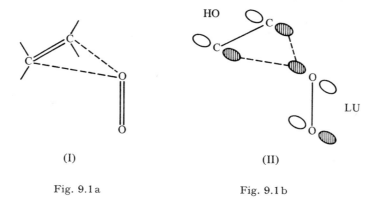

(I)   (II)

Fig. 9.1 a   Fig. 9.1 b

(III)

Fig. 9.1 c

Fig. 9.1. Mode of orbital interaction in the interaction of singlet oxygen with ethylene

For explaining the new idea of "paradoxical" LU—LU orbital control, we consider a system consisting of a donor A and an acceptor B interacting with each other. It is convenient to apply the theory mentioned in Chap. 3 to this problem. Assume the mixing of an excited configuration ψ_K into the zeroconfiguration ψ_0 through a transferred configuration ψ_L.

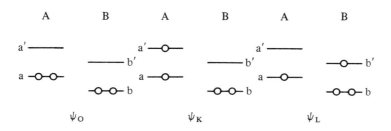

The interaction of these three configurations is described by the third-order term of Eq. (3.9):

$$2 \mathscr{R} \, \frac{(H_{0K} - S_{0K}H_{00}) \, (H_{L0} - S_{L0}H_{00}) \, (H_{KL} - S_{KL}H_{00})}{(H_{00} - H_{KK}) \, (H_{00} - H_{LL})} \tag{9.1}$$

in which $\mathscr{R}$ signifies the real part. The coefficient of ψ_K in the perturbed wave function of Eq. (3.7) is

$$\frac{H_{K0} - S_{K0}H_{00}}{H_{00} - H_{KK}} + \frac{(H_{L0} - S_{L0}H_{00}) \, (H_{KL} - S_{KL}H_{00})}{(H_{00} - H_{LL}) \, (H_{00} - H_{KK})} \tag{9.2}$$

where the first term comes from the second-order term of Eq. (3.9) and corresponds to the *direct* interaction of configurations 0 and K. The second term corresponds to the third-order energy (9.1) and represents the effect of mixing of ψ_K through intermediation of ψ_L. In respect of these two terms the comparison of the order of magnitude is of interest. In the extreme case where the acceptability of B is so large that the energy of configuration ψ_L is equal to that of ψ_0, the second term of Eq. (9.2) is written

$$\frac{H_{L0} - S_{L0}H_{00}}{|H_{0L} - S_{0L}H_{00}|} \cdot \frac{H_{KL} - S_{KL}H_{00}}{H_{00} - H_{KK}}$$

The order of magnitude of this term is $\sim s_{a'b'}$ which implies the overlap integral between MO's a' and b', i.e. LU of A and LU of B, respectively. This obviously overcomes that of the second order term of (9.2) which is $\sim s_{ab}s_{a'b}[60)]$.

The same thing can be said with respect to the energy of the system. The order of magnitude of Eq. (9.1) for the extreme case of $H_{LL} \sim H_{00}$ is $\sim s_{ab}s_{a'b}s_{a'b'}$ which exceeds the second-order term of Eq. (3.9)

$$|H_{0K} - S_{0K}H_{00}|^2 / (H_{00} - H_{KK})$$

of the order $\sim (s_{ab}s_{a'b})^2$. This means that the stabilization of the system is largely controlled by $s_{a'b'}$ that is the LU—LU overlapping in the quasi-four-center interaction illustrated above. A similar paradoxical HO—HO orbital control is also considered.

Therefore, it is possible that the interaction of a strong electron-acceptor enhances the mixing of a locally excited configuration of the donor, or the interaction of a strong donor effectuates the mixing of a locally excited configuration of the acceptor, both of which are otherwise negligibly small. The mixing of such an "excited" configuration frequently brings about unusual reactivity as the result of paradoxical LU—LU or HO—HO orbital control. Such a mechanism may be called *"pseudoexcitation"* [183].

The concept of pseudoexcitation can be applied also to the problems of regioselectivity in four-membered cycloadditions, and thermal and photoinduced reactions through charge-transfer complexation [183].

10. Three-species Interaction

The expression of the third-order term of Eq. (3.9), i.e. Eq. (9.1), finds an interesting application. This term can be negative and therefore contribute to stabilizing the system only if

$$S_{K0}S_{0L}S_{LK} > 0 \qquad (10.1)$$

since for a sufficiently small interaction

$$S_{pq}(H_{qp} - S_{qp}H_{00}) < 0$$

in the argument using real functions. This stability condition (10.1), applied to a system composed of three species interacting with one another, serves frequently as the selection rule of chemical interactions.

Let three species be A, B, and C. We assume the interaction of the following three configurations to be the most important term among the third-order ones. This implies that essentially A is a donor while B and C are acceptors:

The condition Eq. (10.1) becomes

$$S_{ab'}S_{ac'}S_{b'c'} > 0 \qquad (10.2)$$

which represents the following phase relationship in orbital overlappings:

$$(HOMO)_A$$

$$\overset{\oplus}{} \qquad \overset{\oplus}{}$$

$$(LUMO)_B \underset{\oplus}{\underline{\qquad}} (LUMO)_C \qquad (10.3)$$

where $\oplus$ designates an in-phase overlapping.

Similar consideration with regard to the interaction of two donors and one acceptor leads to the condition of stabilization

$$S_{ac}{}' S_{bc}{}' S_{ab} < 0 \qquad (10.4)$$

and accordingly the orbital phase relation

$$(HOMO)_A \overset{\ominus}{\underline{\qquad}} (HOMO)_B$$

$$\overset{\oplus}{} \qquad \overset{\oplus}{}$$

$$(LUMO)_C \qquad (10.5)$$

in which $\ominus$ means an out-of-phase overlapping.

Various illustrations for these results can be given (Fig. 10.1). No explanation will be needed.

Berson[186] extended his study on the bicycloheptene rearrangement (Sect. 7.5) to rather more complicated cases. These provide materials to discuss the three-species interaction.

i) decarboxylation of β-, γ-unsaturated carboxylic acid

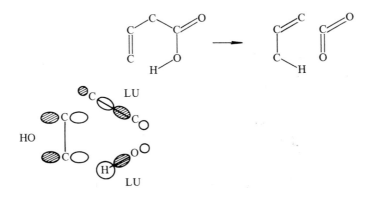

Three-species Interaction

ii) "ene" reaction

iii) Cope rearrangement

iv) "retroene" reaction

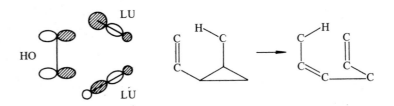

Fig. 10.1. Various three-species interactions

a) Methylenebicycloheptene rearrangement

I: 1,3-alkyl migration

II: Cope rearrangement (Fig. 10.2)

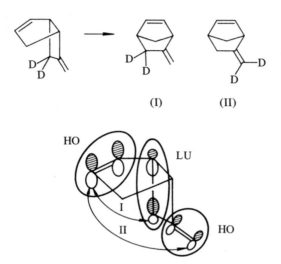

Fig. 10.2. Cope rearrangement

b) Methylenebicyclooctene rearrangement (Fig. 10.3)

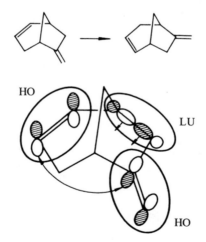

Fig. 10.3. Methylenebicyclooctene rearrangement

91

Another interesting example is the decomposition of alkylchloro-sulfite to form alkyl chloride.

$$ROSOCl \longrightarrow R^+\text{-----}OSOCl^- \longrightarrow RCl + SO_2$$

in which stereochemistry is retained [187]. The contribution of a three-species interaction to stabilizing the transition state can be imagined.

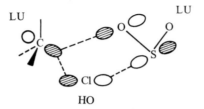

Similar stereospecificity is observed in the SO_2 insertion reaction

$$C_5H_5Fe(CO)_2R + SO_2 \longrightarrow C_5H_5Fe(CO)_2SO_2R \quad [188]$$

The stereochemistry is retained at Fe and inverted at R.

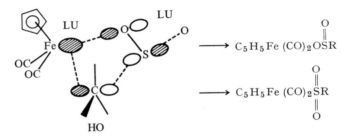

The three-species interaction also contributes to the intramolecular stability.

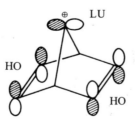

7-norbornadienyl cation

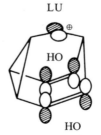

barbaralyl cation

11. Orbital Catalysis

The orbital-interaction scheme developed in the preceding Chapter is conveniently applied to understanding the role of a certain group of catalysts [189]. The mode of interaction of a catalyst C with the reacting system of two species A and B belongs to either one of the following two categories:

I.   A------B-------C

II.   A------B
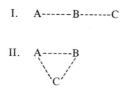

Scheme I represents the case where the catalyst exerts influence upon only one component B, while in Scheme II it affects both of the two components A and B.

Suppose that C in Scheme I is a strong donor (i) or acceptor (ii). The mixing of the electron configuration of B like

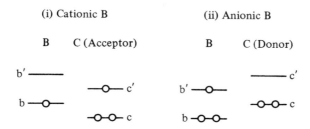

|(i) Cationic B | (ii) Anionic B |

may permit a $(HOMO)_A$—$(HOMO)_B$ or $(LUMO)_A$—$(LUMO)_B$ interaction resulting in occurrence of a reaction which is originally not favorable.

Also the pseudoexcitation mechanism (Chap. 9) may be considered in the catalytic activation of B.

The mixing of these configurations may cause unusual reactivity similar to that of an excited state.

In actual working conditions, the Scheme I activation of reactant *B* may be based on electron transfer or pseudoexcitation mentioned above, or most presumably both mechanisms cooperating. It is difficult to distinguish between these two. Several examples of Scheme I-activation will follow.

1. Opening of unsaturated rings (Fig. 11.1) [190].

   The reaction

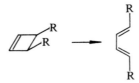

takes place in 40 min in the presence of $AgBF_4$ whereas no reaction occurs after 24 h without catalysts.

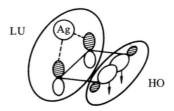

Fig. 11.1. Catalyzed opening of cyclobutenes

2. Simultaneous opening of four-membered ring (Fig. 11.2).

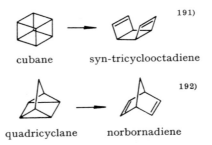

cubane   syn-tricyclooctadiene [191]

quadricyclane   norbornadiene [192]

94

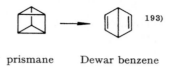

prismane   Dewar benzene   193)

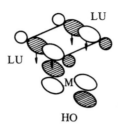

Fig. 11.2. Metal calalyzed opening of cyclobutanes

3. Isomerization of multicyclic saturated hydrocarbons (Fig. 11.3) [194].

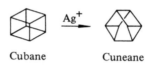

Cubane   Cuneane

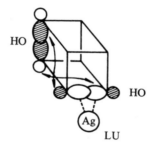

Fig. 11.3. Isomerization of cubane

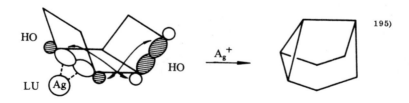

syn-tricyclo [4.2.0.0^{2,5}] octane   tetrahydrosemibullvalene   195)

If the molecule B is molecular oxygen ($^3\Sigma_g^-$) the effect of C will be in the following two directions:

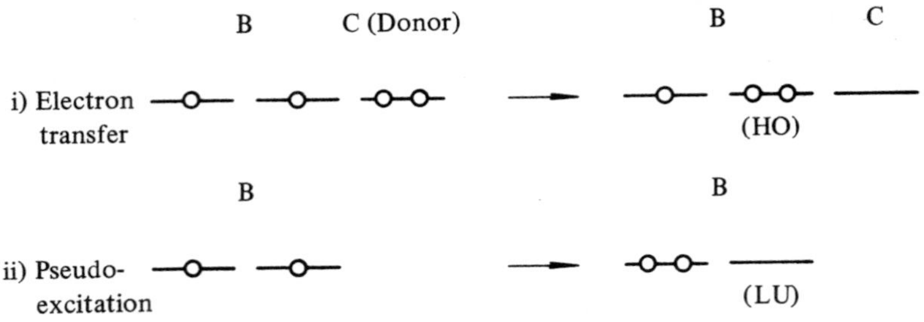

Thus, oxygen molecule in combination with a strong donor behaves as a superoxide (O_2^-)-like reactivity (i).

196)

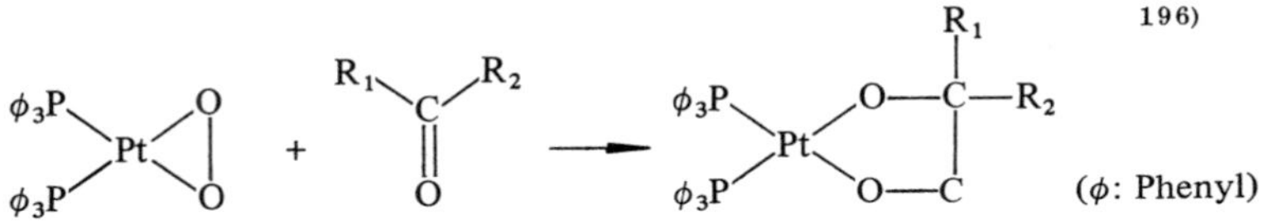

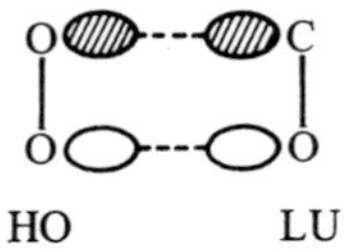

(ϕ: Phenyl)

Under the influence of a certain environment, oxygen molecule undergoes considerable pseudoexcitation (ii) and behaves like singlet excited state ($^1\Delta_g$).

197)

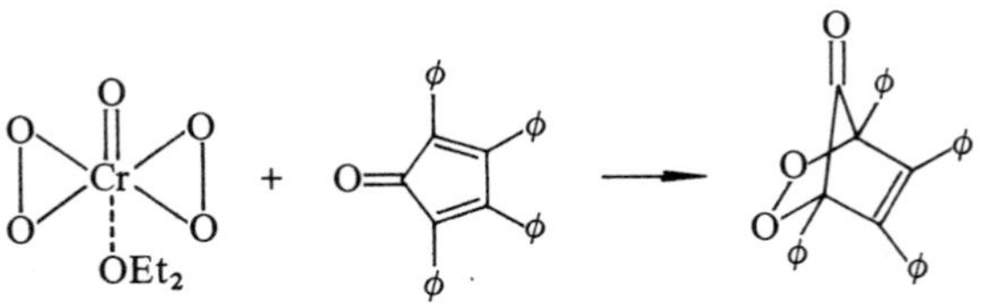

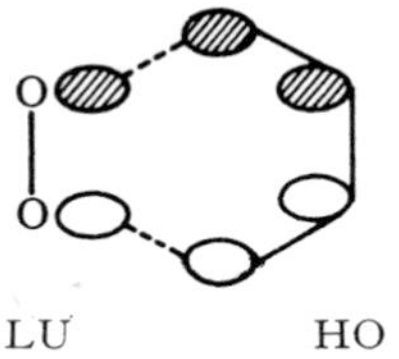

Molecular oxygen oxidizes hydrocarbons in the presence of onium salts [198].

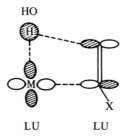

Such an activation mechanism of molecular oxygen seems to take part in biological oxygenation reactions.

It seems likely that the Scheme II-mechanism of activation is frequently more probable and more important than Scheme I in transition metal catalysis. The following examples are those which can conveniently be explained by the use of Scheme II.

1. Insertion (Fig. 11.4).

$$(\pi - C_5 H_5)_2 \, MoH_2 + CH_2 = CHX \longrightarrow (\pi - C_5 H_5)_2 \, MoH - CHXCH_3 \; [199]$$
$$(X = CN, COOCH_3)$$

HO

LU   LU

Fig. 11.4. Insertion of ligands

2. 1.3-hydrogen shift (Fig. 11.5) [200].

$$\varnothing \, CH_2 - CH = CH_2 \xrightarrow{\text{M}} \phi CH = CH - CH_3$$

3-phenyl propene   1-phenyl propene
$$(M = Fe(CO)_5, \; DCo(CO)_4^-)$$

97

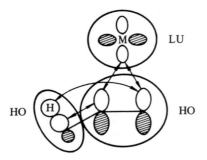

Fig. 11.5. 1,3-hydrogen shift

3. Disproportionation and dimerization of olefins. (Fig. 11.6).

$$R_1R_1C{=}CR_1R_1 + R_2R_2C{=}CR_2R_2 \longrightarrow 2\,R_1R_1C{=}CR_2R_2\ [201)$$

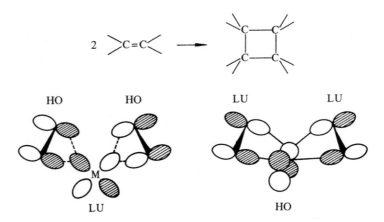

Fig. 11.6. Disproportionation of olefins

4. Recombination of ligands.

[202)]

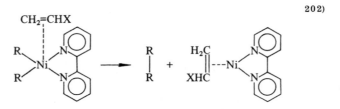

(X: electron-attracting group)

98

12. Thermolytic Generation of Excited States

When a and b are the two different orbitals orthogonal to each other belonging to one molecule, the energy expression, Eq. (8.1), derived in Chap. 8 becomes

$$^1W = h_{aa} + h_{bb} + (aa|bb) + (ab|ab)$$
$$^3W = h_{aa} + h_{bb} + (aa|bb) - (ab|ab)$$

(12.1)

and obviously the triplet state is more stable than the singlet state. The two orbitals happen to have an equal energy, that is a degenerate case.

$$a \longrightarrow\!\!\text{O}\!\!\longrightarrow \quad b \longrightarrow\!\!\text{O}\!\!\longrightarrow$$

Such a case could appear near the HOMO—LUMO crossing [203] of a "forbidden" reaction discussed in Chap. 1 (Fig. 12.1). Mixing of other MO's near the crossing point smoothes the energy levels.

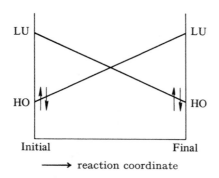

Fig. 12.1. The HOMO—LUMO crossing

The initial singlet state becomes unstable by the existence of a more stable triplet state, which is known as "triplet instability". If in this system the transition of the spin state from singlet to triplet is permitted, it is possible that the product in a triplet state is formed by a thermal process.

1,2-Dioxetanes decompose thermally to produce triplet acetone[204].

$$\begin{array}{c} \text{O}-\text{O} \\ |\quad| \\ \text{Me}-\text{C}-\text{C}-\text{Me} \\ |\quad| \\ \text{Me}\ \ \text{Me} \end{array} \longrightarrow [(CH_3)_2CO]^* + (CH_3)_2CO \\ \text{(triplet)}$$

The orbital interaction rationale for this process is as follows:

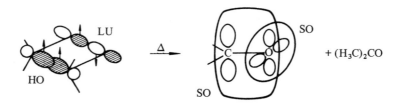

The HOMO—LUMO interaction results in the decomposition of the direction indicated by arrows. In the acetone molecule to be formed on the right-hand side of the figure, the two odd-electron orbitals tend to point to the directions perpendicular to each other, so as to fit the energetical condition necessary for making the triplet state more stable.

The acetone molecule thus formed has an electronic structure similar to n—π^* excitation. Dewar has shown theoretically that the crossing of singlet and triplet states occurs before the transition state and one acetone molecule to be formed can be in the lowest excited triplet state [205].

Bicyclo[2.2.0]hexadiene (Dewar benzene) produces excited triplet benzene through thermal ring opening [206].

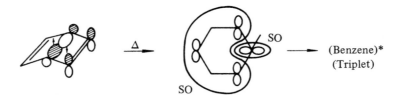

Dewar calculated the path of this process and found that the singlet-triplet crossing takes place after the transition state and accordingly the triplet generation is in this case inefficient [207]. .

The thermolytic formation of excited state molecules of this kind has a close connection with the theory of chemiluminescence [208].

The essential process of photoemission of luciferin-luciferase system [209] is considered as the decomposition of dioxetanone ring.

(n–π*–Like excited state)

One can see in Eq. (12.1) that the singlet-triplet separation is essentially attributed to the repulsion integral

$$(n\pi^*|n\pi^*)$$

where n denotes the oxygen lone-pair orbital and π^* the LUMO of the conjugated part. Hence, it is expected that the delocalization of π^* and n orbitals will decrease the magnitude of this integral. If the n-π^* excited ketone assumed to be produced by the dioxetanone ring opening has a chromophore which can conjugate with the carbonyl group, the singlet-triplet separation will be reduced. Also the existence of the lone-pair electrons on the neighboring nitrogen atom will cause the delocalization of odd electron in the n orbital to assist in reducing the repulsion integral. In such cases it may be possible that the excited state generated is singlet.

13. Reaction Coordinate Formalism

The principles governing the geometrical change of the reacting species along the reaction path have been discussed in Chap. 4 from the standpoint of orbital interaction. A more general treatment is here desired to image a fundamental picture of chemical reactions.

The reaction coordinate can be defined, on the potential surface of the reacting system in the frame of the Born-Oppenheimer approximation, as the path along which the geometry of the reacting system changes at an infinitely slow speed.

Let the generalized coordinates to determines the nuclear location be $\xi_1, \xi_2 \ldots \xi_n$. The Lagrangian can be written as

$$L = \frac{1}{2} \sum_{i,j}' a_{ij} \frac{d\xi_i}{dt} \frac{d\xi_j}{dt} - W(\xi_1, \xi_2, \ldots \xi_n)$$

where W is the potential energy and $a_{ij} (= a_{ji})$'s are obtained as functions of ξ_i's from their relation with the Cartesian coordinates. The Equations of nuclear motion are

$$\frac{d}{dt} \sum_j' a_{ij} \frac{d\xi_j}{dt} = -\frac{\partial W}{\partial \xi_i} \quad (i = 1, 2, \ldots . n)$$

Integrating these Equations within a small time interval under the assumption that $d\xi_i/dt = 0$ at $t = 0$ by definition, one obtains

$$\sum_j' a_{ij} \frac{d\xi_j}{dt} = \left(-\frac{\partial W}{\partial \xi_i}\right) t \quad (i = 1, 2, \ldots . n)$$

which leads at once to

$$\frac{\sum_i a_{1i} d\xi_i}{\dfrac{\partial W}{\partial \xi_1}} = \frac{\sum_i a_{2i} d\xi_i}{\dfrac{\partial W}{\partial \xi_2}} = \ldots \ldots = \frac{\sum_i a_{ni} d\xi_i}{\dfrac{\partial W}{\partial \xi_n}} \tag{13.1}$$

This relation holds from moment to moment on the reaction co-ordinate. If a set of initial values $(\xi_{10}, \xi_{20}, \ldots \xi_{n0})$ is given, the reaction coordinate is obtained as a function of one parameter by solving this differential equation.

When Cartesian coordinates $(X_\alpha, Y_\alpha, Z_\alpha)$ of nucleus α are used, Eq. (13.1) becomes

$$\ldots = \frac{M_\alpha dX_\alpha}{\dfrac{\partial W}{\partial X_\alpha}} = \frac{M_\alpha dY_\alpha}{\dfrac{\partial W}{\partial Y_\alpha}} = \frac{M_\alpha dZ_\alpha}{\dfrac{\partial W}{\partial Z_\alpha}} = \ldots \qquad (13.2)$$

in which M_α is the mass of nucleus α.

Eq. (13.2) indicates that the direction of displacement of nuclear location conforms to that of the gradient of potential energy. Accordingly, the reaction coordinate belongs to the same symmetry as the nuclear arrangement. The symmetry is thus conserved in the deformation along the reaction coordinate [210].

In an equilibrium point, where all of the gradients of W vanish, the reaction coordinate of Eq. (13.1) coincides with a normal coordinate [210]. Therefore, one can learn the character of a reaction coordinate by scrutinizing the corresponding normal coordinate [211].

Reaction   Normal coordinate

1. Decomposition of carbon dioxide [211].

$$CO_2 \longrightarrow O + CO \; (D_{\infty h}) \qquad \text{←O—C—►O} \qquad \Sigma_u^+$$

2. Decomposition of the transition state of S_N2 reaction [211].

3. Decomposition of cyclobutane [212].

The symbol of the right end signifies the symmetry of the normal coordinate, and that in the bracket is the symmetry of the molecule.

It is shown that at an equilibrium point the direction of the normal vibration which possesses the same symmetry as the product of HOMO

103

and LUMO of the molecule corresponds to that of a favorable reaction coordinate [211]. The reactions mentioned above (1, 2, 3) are the examples.

At a non-equilibrium point the reaction coordinate on which nuclei move to the region of HOMO—LUMO overlapping corresponds to a favorable path [213]. Therefore, for a reaction path to be a favorable one, HOMO and LUMO have to belong to the same symmetry [214]. This conclusion is nothing but the criterion of the favorable direction of reactions already given in Chap. 5.

The importance of HOMO—LUMO interaction on the reaction coordinate was tested in the following several cases:

1. Bimolecular nucleophilic replacement [215].
2. Dimerization of methylenes [216].
3. Addition of singlet methylene to ethylene [217].
4. Abstraction of methane hydrogen by methyl [218].
5. Addition of methyl to ethylene [218].

An analysis of forces acting on nuclei along the reaction path was made with respect to the hydrogen substitution of methane by tritium

$$T + CH_4 \longrightarrow TCH_3 + H \text{ [219]}$$

In this case, exchange force and Coulomb force obstruct the proceeding of reaction. The force accompanying the electron delocalization, that is, the mutual charge-transfer interaction, between methane and tritium acts to promote the reaction. In particular, the most important is the delocalization force between HOMO and LUMO of CH_4 and T.

14. Correlation Diagram Approach

The use of the MO correlation diagram for discussing the electron configuration of diatomic molecules is well established [220]. The Walsh rule [85], based on the correlation of MO's with a deformation parameter in simple molecules, is conveniently employed to guess the stable geometrical form. The fundamental idea utilized in constructing these diagrams is the application of the "noncrossing rule" [221] which states that two MO's with the same symmetry do not cross.

The reaction coordinate has a definite symmetry throughout, as mentioned in Chap. 13. Accordingly, by group theoretical requirement, the MO symmetry is conserved along the reaction coordinate throughout. Therefore, one can draw an MO correlation diagram with the reaction coordinate varied by MO symmetry argument.

It is worth special mention that this type of correlation diagram was applied to discussing stereoselective phenomena by Woodward and Hoffmann with great success [222]. Experimental results of various stereoselective reactions were thoroughly collected and classified into several fundamental types with the use of new terminology. Such a classification is well comprehensible to organic chemists and is widely used.

The criterion for a "forbidden" and an "allowed" reaction is essentially based on whether the HOMO-LUMO crossing, or more rigorously the occupied-unoccupied MO correlation, does or does not occur in the MO correlation diagram of the reacting system (Fig. 14.1).

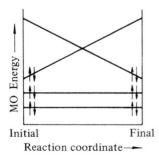
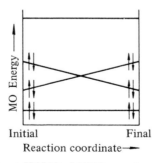

i) With HOMO—LUMO crossing   ii) Without HOMO—LUMO crossing

Fig. 14.1. Correlation diagram for the reacting system

One should consult the original paper of Woodward and Hoffmann which gave an excellent comprehensive explanation for this criterion. Only one example of the thermal ring-opening of cyclobutene already mentioned in Sect. 7.5 is cited here in Fig. 14.2. A HOMO—LUMO correlation is seen in the "forbidden" disrotatory path.

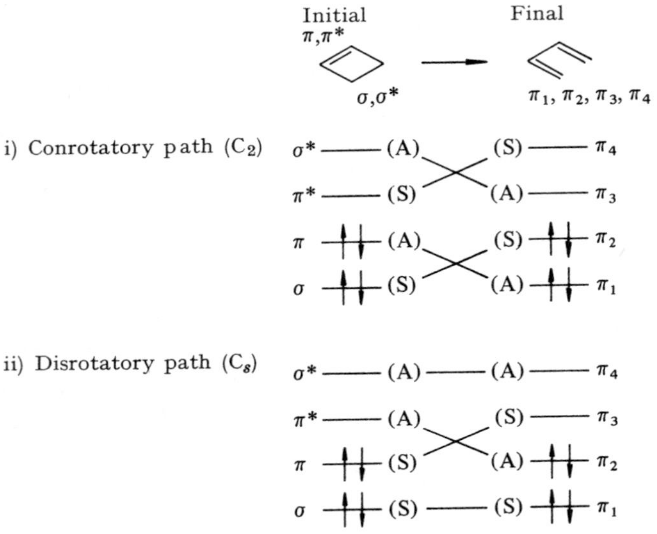

Fig. 14.2. Correlation diagram for the ring-opening of cyclobutene

$(C_2, C_s:$ symmetry of the reaction coordinate)
$(S, A:$ symmetric and antisymmetric)

The occupied-unoccupied MO crossing or noncrossing is no doubt a helpful criterion. However, if one wants to rely upon this criterion only, a certain inconsistency could arise. For instance, consider a reaction coordinate which is symmetric with respect to time reversal:

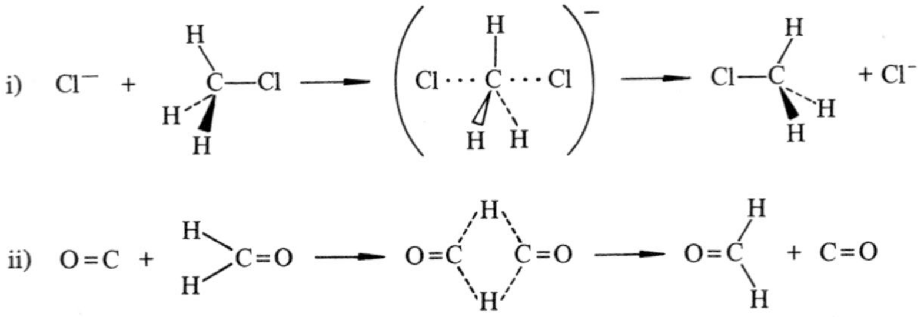

Evidently, the reaction (ii) is hypothetical and very unlikely to take place, and we can imagine many other similar cases. Consider an arbitrary case of such a reaction. When there is no occupied-unoccupied correlation, there will be no problem in putting this reaction into the "allowed" group. And suppose tentatively that, as is in the reaction of example (i) [211], the HO (or a certain occupied) MO of the initial system happens to go into the LU (or a certain unoccupied) MO of the final system in the correlation diagram of this type of reactions. Then these two MO's must split into two different levels near the midpoint of the diagram by the noncrossing rule, since these two MO's must belong to the same symmetry type of the point group to which the reaction coordinate belongs.

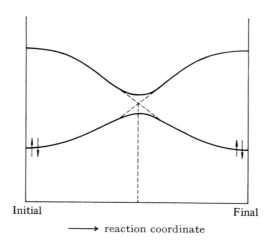

Initial   Final

⟶ reaction coordinate

Therefore, all such reactions come to be classified into the "allowed" group, however large the activation energy might be, so far as one adopts the occupied-occupied or occupied-unoccupied correlation as the absolute criterion for judging the favorableness or unfavorableness of the reaction.

On the other hand, if one puts all this type of reactions, where a certain occupied orbital level has a maximum of similar origin in the correlation diagram, into the "forbidden" class, the first example, the common S_N2 reaction, becomes "forbidden". Thus, the occurrence or non-occurrence of occupied-unoccupied correlation can be no measure of the magnitude of activation energy. And, if we have to take the magnitude of activation energy, large or small, as the scale of our judgement, we are compelled to say that there would be little advantage in using the correlation diagram drawn qualitatively by the aid of non-crossing rule or by a simple symmetry argument.

The orbital interaction approach hitherto mentioned is responsible for interpreting why an occupied-unoccupied crossing does or does not take place in the correlation diagram by the stabilization-energy argument of the reacting system. It is easily understood that the example (i) mentioned above can actually be an allowed reaction since the HOMO-LUMO separation of the initial system is so narrow that their crossing affects the activation energy only slightly.

15. The Nature of Chemical Reactions

As is understood from the results mentioned in preceding Chapters, the orbital interaction approach has the following advantages:

1. The favorableness of a reaction path is directly estimated by the HOMO—LUMO criteria of reactants.

2. The reactivity, orientation, and stereoselection of various reactions, inter- and intramolecular, one-centre and multicentre, ground- and excited-state, like substitution, abstraction, addition, elimination, disproportionation, isomerization, thermolytic formation of excited states, and so on, and also catalysis, molecular stability, intermolecular interaction, or molecular shape, of all sorts of compounds, organic or inorganic, saturated or unsaturated, can be discussed uniformly.

3. The same way of thinking can be applied to every case without introduction of particular concepts or new terminology with respect to the type of reaction, where neither symmetric property nor "selection rule" is required.

4. The continuity of selection rule and that of concertedness, the effect of neighboring groups, including regioselectivity and periselectivity, that of conjugating groups or substituents, that of interacting species like solvents, the singlet-triplet selectivity, and other secondary effects are conveniently interpreted.

5. The molecular mechanism of the course of a reaction is automatically given by the analysis of forces acting on nuclei along the reaction pathway.

The success of the *intermolecular orbital interaction picture* thus enables the discussion of the nature of chemical reactions.

It is easily understood by the use of the usual perturbation theory for two interacting molecules that the interaction between one of occupied MO's of one molecule and one of occupied MO's of the other ("occupied-occupied" interaction) contributes nothing to stabilize the interacting system nor to form a new bond between the two molecules. It is the "occupied-unoccupied" interaction that can produce stabilization of the system and accumulate electrons between the two molecules to form a new bond. Among occupied-unoccupied orbital interactions, the electron delocalization due to the HOMO—LUMO overlapping is usually most important since a weak orbital interaction is proportional to the orbital

109

overlapping and inversely proportional to the level separation. The orientation in the initial stage of interaction is thus essentially governed by the HOMO—LUMO overlapping.

This conclusion is ascertained, by actual analysis of the interaction energy along the reaction path.

As has been discussed in Chap. 3, the chemical interaction between molecular systems is divided into the Coulomb interaction, the exchange interaction, the electron delocalization, and the polarization interaction. By way of the electron delocalization and the polarization interaction, the different electron configurations come to mix with the initial one. The molecular shape will tend to change so as to conform to this new electron distribution. That is to say, the change in the electron distribution impels the nuclei to rearrange themselves.

Of the two sorts of interaction mentioned above, electron delocalization is shown to be more important than polarization. One of the reactants happens to be an electron-donor and the other an electron-acceptor, as is the case in most heterolytic reactions. In such cases the electron delocalization obviously predominates over the polarization effect. Even in homolytic interactions, the importance of the mutual delocalization is shown to be remarkable. In the majority of actual cases of the *electron delocalization* interaction in which reactants are free to change their nuclear configuration, the HOMO—LUMO delocalization is calculated to be dominant.

As the reaction proceeds, the HOMO and the LUMO come to localize more at the reaction centers and the HOMO—LUMO level separation becomes narrower, resulting obviously in making the role of these particular MO's more important. The conspicuous role of the HOMO—LUMO interaction is thus more distinct, as has been discussed in Chap. 4 (and in Appendix I). The electron delocalization from HOMO to LUMO between two reactant molecules contributes to the bond exchange necessary for the occurrence of the reaction. Such nuclear rearrangement is accompanied by a destabilization, like promotion in molecule formation from atoms, which is the principal origin of "activation energy", and the electron delocalization helps the reaction to proceed smoothly, in which the particular orbitals, HOMO and LUMO of both reactants, play the principal role.

This is the probable mechanism of determination of the reaction pathway. In the reactions in which a radical or an excited molecule takes part, the overlapping of the SOMO of the specie with the other reactant is of significance and plays the part of HOMO or LUMO of common molecules.

The ease with which the reaction proceeds is directly related to the property or behavior of these particular MO's connecting these to the phenomena of orientation or stereoselection. The electron distribution (valence-inactive population) plays a leading role in the interaction between the particular orbitals, HOMO, LUMO, and SOMO, in usual molecules, and determines the orientation in the molecule in the case of chemical interaction. In that case, the extension and the nodal property of these particular MO's decide the spatial direction of occurrence of interaction.

It is understood that the direct "motive force" which drives a sizeable molecule, even a complicated organic molecule, to chemical reaction may be ascribed to the behavior of electrons whose mass is less than a ten- or hundred-thousandth of that of the molecule. In some cases the existence of a field of "vacant" orbitals extending for long distances facilitates the initial interaction and gives the reagent a chance to select the reaction path.

The nature of intermolecular force is not essentially different from that which participates in the chemical bond or chemical reaction. The factor which determines the stable shape of a molecule, the influence on the reaction of an atom or group which does not take any direct part in the reaction, and various other sterically controlling factors might also be comprehended by a consideration based on the same theoretical foundation. The effect of solvents, and the transfer of energy from one molecule to another, are also the object of discussion in the same category.

The secondary or tertiary structure of high polymers, the catalytic effect of organic molecules in the majority of enzymatic reactions, and common chemical interactions in heterogeneous systems may also be controlled by factors of the same category. The high selectivity observed in these sorts of interaction might originate from the selectivity of the molecular field which is formed by the complicated molecular systems involved.

The possibility afforded by such a molecular field to possess any cause of selectivity in the chemical interaction will easily be recognized by the previous discussions.

Appendix I. Principles Governing the Reaction Path

An MO-Theoretical Interpretation

For the purpose of discussing simply the essential feature of the principles governing the reaction path, it is convenient to apply the extended Hückel method. The following relations hold among the orbital energy ε_i, the coefficient of r th AO of i th MO c_{ir}, the Coulomb integral h_{rr}, and the resonance integral h_{rs}:

$$(h_{rr} - \varepsilon_i)c_{ir} + \sum_{s(\neq r)} (h_{rs} - s_{rs}\varepsilon_i)c_{is} = 0 \qquad \text{(AI-1)}$$

where s_{rs} is the overlap integral and

$$h_{rs} = \frac{K}{2}(h_{rr} + h_{ss})s_{rs} \qquad \text{(AI-2)}$$

in which K is a constant larger than unity. For an extremely rough estimation, all the Coulomb integrals are assumed constant and equal to $\bar{h}$. We can easily obtain

$$\frac{v_{ir}}{p_{ir}} \sim \frac{\varepsilon_i - \bar{h}}{K\bar{h} - \varepsilon_i} \qquad \text{(AI-3)}$$

where

$$v_{ir} = \sum_{s(\neq r)} c_{ir}c_{is}s_{rs} = n_{ir} - p_{ir}$$

and

$$p_{ir} = c_{ir}^2$$

are valence-active and valence-inactive parts of AO population n_{ir} [79]. Since in occupied MO's it usually follows that

$$\bar{h} > \varepsilon_i > K\bar{h}$$

we have

$$v_{ir} > 0$$

while in unoccupied MO's

$$v_{ir} < 0$$

since usually the relation

$$\varepsilon_i > h$$

holds. Therefore, by definition $|v_{ir}|$ represents the ease in loosening the bonds between r th AO and the other AO's of the molecule when an electron goes out of an occupied orbital i or goes into an unoccupied orbital i. Then Eq. (AI-3) indicates the parallel relation between $|v_{ir}|$ and p_{ir}. Hence, the position of large amplitude of i th MO is at the same time the position where the contribution of i th MO to the bond-weakening is large. With respect to HOMO and LUMO, the positions of large amplitude determine the reactivity (Chap. 6). Accordingly, *the position of large reactivity is also the position of facile bond exchange in case of electron release or acceptance.*

Next, from (AI-1) it follows that

$$\varepsilon_i = \sum_A^A \sum_r c_{ir}{}^2 h_{rr} + \sum_{(AB)}^{A \ B} \sum_r \sum_s c_{ir} c_{is} h_{rs} \tag{AI-4}$$

The notation $\sum_r^A$, $\sum_A$, and $\sum_{(AB)}$ imply summation over all AO's belonging to atom A, over all atoms, and over all atom pairs, respectively. We have to consider the change in orbital energy $\delta \varepsilon_i$ due to the change of h_{rs}, δh_{rs}, caused by the molecular deformation along actual reaction paths.

$$\delta \varepsilon_i = \sum_{(AB)}^{A \ B} \sum_r \sum_s c_{ir} c_{is} \delta h_{rs}$$

Now, we classify atom pairs into three groups, $(AB)_1$, $(AB)_2$, and $(AB)_3$. In $(AB)_1$ the distance $A-B$ stretches, while that in $(AB)_2$ shrinks. In the remaining group, $(AB)_3$, the $A-B$ distance does not change. Then we have

$$\delta \varepsilon_i \sim \sum_{(AB)_1}^{A \ B} \sum_r \sum_s c_{ir} c_{is} |\delta h_{rs}|$$
$$- \sum_{(AB)_2}^{A \ B} \sum_r \sum_s c_{ir} c_{is} |\delta h_{rs}| \tag{AI-5}$$

113

As mentioned before, the position of large HOMO or LUMO amplitude is also the position where the bonds with neighbors actually weaken in forming a new bond with different molecules. Therefore, the following relation *nearly* holds for *important* (rs) pairs:

i) when MO ψ_i is HOMO $\quad c_{ir}c_{is} > 0$ for $(AB)_1$

$\qquad\qquad\qquad\qquad\qquad c_{ir}c_{is} < 0$ for $(AB)_2$

ii) when MO ψ_i is LUMO $\quad c_{ir}c_{is} < 0$ for $(AB)_1$

$\qquad\qquad\qquad\qquad\qquad c_{ir}c_{is} > 0$ for $(AB)_2$

Accordingly, $\delta\varepsilon_{HO}$ comes to have a large positive value in comparison with other occupied orbitals, $\delta\varepsilon_{LU}$, however, a large negative value in comparison with other unoccupied MO's. Namely, *the HOMO level rises and the LUMO level drops as the bond exchange proceeds in actual reactions.*

Further, we discuss the effect of the weakening of a given bond $A-B$ upon the orbital amplitude and the bond order. Consider an AO r belonging to atom A and an AO s belonging to atom B. The change of MO ψ_i and that of the partial bond order of any AO pair (tu) due to the change of s_{rs}, δs_{rs}, of the AO pair (rs) are given from (AI-1) by

$$\delta\psi_i = -\,c_{ir}c_{is}\psi_i\delta s_{rs} + \sum_{j(\neq i)} \frac{c_{ir}c_{js} + c_{is}c_{jr}}{\varepsilon_i - \varepsilon_j}\,\psi_j(\delta h_{rs} - \varepsilon_i\delta s_{rs}) \qquad \text{(AI-6)}$$

and

$$\delta(c_{it}c_{iu}) = -\,2c_{ir}c_{is}c_{it}c_{iu}\delta s_{rs}$$
$$+ \sum_{j(\neq i)} \frac{(c_{ir}c_{js} + c_{is}c_{jr})(c_{it}c_{ju} + c_{iu}c_{jt})}{\varepsilon_i - \varepsilon_j}\,(\delta h_{rs} - \varepsilon_i\delta s_{rs}) \qquad \text{(AI-7)}$$

In particular, if one puts $t = r$ and $u = s$, this becomes

$$\delta(c_{ir}c_{is}) = -\,2c_{ir}{}^2c_{is}{}^2\delta s_{rs} + \sum_{j(\neq i)} \frac{(c_{ir}c_{js} + c_{is}c_{jr})^2}{\varepsilon_i - \varepsilon_j}\,(\delta h_{rs} - \varepsilon_i\delta s_{rs}) \qquad \text{(AI-8)}$$

If we take the signs of r th and s th AO in such a way that s_{rs} becomes positive, then we have

$$\delta h_{rs} > 0, \quad \delta s_{rs} < 0$$

since "bond" (rs) relaxes by the stretched $A-B$ distance, and hence

$$\delta h_{rs} - \varepsilon_i\delta s_{rs} > 0$$

Here, let MO i be HOMO. The nonbonding orbitals j will not largely contribute to the second term of the right-hand side of (AI-8) compared with the occupied orbitals which give positive terms. Therefore, $\delta(c_{ir}c_{is})$ will be positive. This means an increase in partial bond order. Namely, HOMO becomes more localized at the bond which is bonding in HOMO when it is weakened.

Similarly, if MO i is LUMO, the second term of the right-hand side of (AI-8) is generally negative. The first term is positive in the form as it is. This implies that in nonbonding orbitals the AO *coefficients* have large absolute values owing to negative overlaps and accordingly become small with reduced overlaps. However, the actual *orbital amplitude* is lessened by negative overlaps and therefore extends more by decreasing overlaps. Consequently, the LUMO lobe increases in the bond when it is stretched, which is originally antibonding in LUMO. Thus, *the HOMO and the LUMO become more localized at the bonds according as they are stretched by deformation due to intermolecular electron delocalization between the HOMO and the LUMO of reactant molecules.*

An additional interesting conclusion is as follows. Assume that an AO r has a large HOMO or LUMO amplitude in a molecule and the bond between r and another AO s loosens as the consequence of HOMO–LUMO interaction with another molecule. Applying Eq. (AI-7) to a given pair (rs), summing it over s, then putting $t = u = r$, and dividing it by $2c_{ir}$, we obtain

$$\delta c_{ir} = - \sum_{s\,(\neq r)} c_{ir}{}^2 c_{is} \delta s_{rs} + \sum_{s\,(\neq r)} \sum_{j\,(\neq i)} \frac{c_{ir}(c_{ir}c_{js} + c_{is}c_{jr})}{\varepsilon_i - \varepsilon_j} (\delta h_{rs} - \varepsilon_i \delta s_{rs})$$

Adopting the approximation used in deriving (AI-3) again, i.e.

$$h_{rs} = K\bar{h}s_{rs}$$

and assuming that δs_{rs} is proportional to s_{rs}, namely

$$\delta s_{rs} = - a s_{rs} \quad (a = \text{positive constant})$$

one obtains

$$\delta c_{ir} = - a \frac{\bar{h} - \varepsilon_i}{K\bar{h} - \varepsilon_i} c_{ir}{}^3 + 2a(\bar{h} - \varepsilon_i) c_{ir} \sum_{j\,(\neq i)} \frac{c_{jr}{}^2}{\varepsilon_i - \varepsilon_j}$$

$$+ a(K - 1)\bar{h}c_{ir} \sum_{j\,(\neq i)} \frac{c_{jr}{}^2}{K\bar{h} - \varepsilon_j}$$

Appendix I. Principles Governing the Reaction Path

since

$$\sum_{s\,(\neq r)} c_{ir}S_{rs} = -\frac{\bar{b} - \varepsilon_i}{K\bar{b} - \varepsilon_i}\, c_{ir}$$

by (AI-1). Noting the relation

$$\sum_{j} \frac{c_{jr}{}^2}{K\bar{b} - \varepsilon_j} = \frac{1}{(K-1)\bar{b}} \left(\because \sum_{s} c_{js}S_{rs} = \frac{(K-1)\bar{b}}{K\bar{b} - \varepsilon_j}\, c_{jr} \right)$$

it follows that

$$\delta c_{ir} = a c_{ir} \left\{ 1 - c_{ir}{}^2 + 2(\bar{b} - \varepsilon_i) \sum_{j(\neq i)} \frac{c_{ir}{}^2}{\varepsilon_i - \varepsilon_j} \right\} \qquad \text{(AI-9)}$$

Since we may usually consider that

$$\varepsilon_{HO} < \bar{b} < \varepsilon_{LU}$$

the quantity in the bracket of the right-hand side of Eq. (AI-9) becomes positive if $i =$ HO or LU. In this way, it is shown that the HOMO *and the LUMO become more localized at the reaction center in a molecule in case of formation of a new bond there with a different molecule by the HOMO— LUMO interaction and the subsequent bond-loosening between the reaction center and the remaining part of the molecule.*

Appendix II. Orbital Interaction between Two Molecules

Consider a molecule A subjected to the effect of an outer field due to an approaching molecule B. Let the MO's of A and B be ψ_{Ai} and ψ_{Bk} and their energies ε_{Ai} and ε_{Bk}. The one-electron Hamiltonian of the total system is h, and h'_A is the change of one-electron Hamiltonian of molecule A due to the approach of molecule B, h'_B being the analogous quantity for molecule B. The following integrals are defined for constructing the perturbed secular determinant.

$$s_{ik} = \int \psi_{Ai} \psi_{Bk} dv$$
$$h_{ik} = \int \psi_{Ai} h \psi_{Bk} dv$$
$$h'_{ij} = \int \psi_{Ai} h'_A \psi_{Aj} dv$$
$$h'_{kl} = \int \psi_{Bk} h'_B \psi_{Bl} dv$$

On solving the perturbed secular equation, the perturbed MO's and their energies of A are obtained as

$$
\begin{aligned}
\psi'_{Ai} = \Bigg\{ & 1 - \sum_k^B \frac{s_{ik}(h_{ik} - s_{ik}\varepsilon_{Ai})}{\varepsilon_{Ai} - \varepsilon_{Bk}} - \frac{1}{2} \sum_k^B \frac{(h_{ik} - s_{ik}\varepsilon_{Ai})^2}{(\varepsilon_{Ai} - \varepsilon_{Bk})^2} \\
& - \frac{1}{2} \sum_{j(\neq i)}^A \frac{h'^2_{ij}}{(\varepsilon_{Ai} - \varepsilon_{Aj})^2} \Bigg\} \psi_{Ai} + \sum_{j(\neq i)}^A \Bigg\{ \frac{h'_{ij}}{\varepsilon_{Ai} - \varepsilon_{Aj}} \\
& + \sum_k^B \frac{(h_{ik} - s_{ik}\varepsilon_i)(h_{jk} - s_{jk}\varepsilon_i)}{(\varepsilon_{Ai} - \varepsilon_{Aj})(\varepsilon_{Ai} - \varepsilon_{Bk})} - \frac{h'_{ij}(h'_{ii} - h'_{jj})}{(\varepsilon_{Ai} - \varepsilon_{Aj})^2} \Bigg\} \psi_{Aj} \qquad \text{(AII-1)} \\
& + \sum_k^B \Bigg\{ \frac{h_{ik} - s_{ik}\varepsilon_i}{\varepsilon_{Ai} - \varepsilon_{Bk}} + \sum_{j(\neq i)}^A \frac{h'_{ij}(h_{jk} - s_{jk}\varepsilon_i)}{(\varepsilon_{Ai} - \varepsilon_{Aj})(\varepsilon_{Ai} - \varepsilon_{Bk})} - \frac{s_{ik}h'_{ii}}{\varepsilon_{Ai} - \varepsilon_{Bk}} \\
& + \sum_{l(\neq k)}^B \frac{h'_{kl}(h_{il} - s_{il}\varepsilon_i)}{(\varepsilon_{Ai} - \varepsilon_{Bk})(\varepsilon_{Ai} - \varepsilon_{Bl})} - \frac{(h_{ik} - s_{ik}\varepsilon_i)(h'_{ii} - h'_{kk})}{(\varepsilon_{Ai} - \varepsilon_{Bk})^2} \Bigg\} \psi_{Bk} \\
& + 0(\varDelta^3)
\end{aligned}
$$

and

$$\varepsilon'_{Ai} = \varepsilon_i + h'_{ii} + \sum_{k}^{B} \frac{(h_{ik} - s_{ik}\varepsilon_i)^2}{\varepsilon_{Ai} - \varepsilon_{Bk}} + \sum_{j(\neq i)}^{A} \frac{h'^2_{ij}}{\varepsilon_{Ai} - \varepsilon_{Aj}}$$

$$+ 2 \sum_{\substack{j<m \\ (j,m \neq i)}}^{A} \frac{h'_{ij}h'_{im}h'_{jm}}{(\varepsilon_{Ai} - \varepsilon_{Aj})(\varepsilon_{Ai} - \varepsilon_{Am})}$$

$$+ 2 \sum_{j(\neq i)}^{A} \sum_{k}^{B} \frac{h'_{ij}(h_{ik} - s_{ik}\varepsilon_i)(h_{jk} - s_{jk}\varepsilon_i)}{(\varepsilon_{Ai} - \varepsilon_{Aj})(\varepsilon_{Ai} - \varepsilon_{Bk})} \qquad \text{(AII-2)}$$

$$+ 2 \sum_{k<l}^{B} \frac{h'_{kl}(h_{ik} - s_{ik}\varepsilon_i)(\varepsilon_{il} - s_{il}\varepsilon_i)}{(\varepsilon_{Ai} - \varepsilon_{Bk})\cdot(\varepsilon_{Ai} - \varepsilon_{Bl})} - 2 \sum_{k}^{B} \frac{s_{ik}h'_{ii}(h_{ik} - s_{ik}\varepsilon_i)}{\varepsilon_{Ai} - \varepsilon_{Bk}}$$

$$- \sum_{j(\neq i)}^{A} \frac{h'^2_{ij}(h'_{ii} - h'_{jj})}{(\varepsilon_{Ai} - \varepsilon_{Aj})^2} \sum_{k}^{B} \frac{(h'_{ii} - h'_{kk})(h_{ik} - s_{ik}\varepsilon_i)^2}{(\varepsilon_{Ai} - \varepsilon_{Bk})^2}$$

$$+ 0\,(\Delta^4)$$

where Δ signifies the first-order quantities like s_{ik}, h_{ik}, h'_{ij}, h'_{kl}, and so on, and $0(\Delta^3)$ implies a small quantity of order Δ^3 [223].

An important consequence of these Equations may be the rule of intramolecular orbital mixing by the orbitals of a different molecule, which is given by the coefficient of ψ_{Aj} in ψ'_{Ai}. This is written as

$$\frac{h'_{ij}}{\varepsilon_{Ai} - \varepsilon_{Aj}} \sum_{k}^{B} \frac{(h_{ik} - s_{ik}\varepsilon_i)(h_{jk} - s_{jk}\varepsilon_i)}{(\varepsilon_{Ai} - \varepsilon_{Aj})(\varepsilon_{Ai} - \varepsilon_{Bk})} \qquad \text{(AII-3)}$$

if we neglect the remaining term as less important than these two. The first term originates from the "static" effect of the field of molecule B upon the Hamiltonian for molecule A, while the second is caused by the "dynamic" effect of the orbitals of the second molecule [224]. Both constitute the effect of polarization of molecule A.

Next, we take only the direct, "dynamic" orbital effect among three orbitals ψ_{Ai}, ψ_{Aj}, and ψ_{Bk} into account, and consider the phase relation in the perturbed orbital ψ'_{Ai} represented by

$$\psi'_{Ai} = \psi_{Ai} + \frac{h_{ik}h_{jk}}{(\varepsilon_{Ai} - \varepsilon_{Aj})(\varepsilon_{Ai} - \varepsilon_{Bk})}\,\psi_{Aj} + \frac{h_{ik}}{\varepsilon_{Ai} - \varepsilon_{Bk}}\,\psi_{Bk} \qquad \text{(AII-4)}$$

where the overlap integrals are neglected as small. The following four cases arise in respect to the orbital level relation.

The resulting ψ'_{Ai} is given by the following Table:

$$\psi'_{Ai}$$

i) $\varepsilon_{Ai} > \varepsilon_{Ai}, \varepsilon_{Bk} > \varepsilon_{Ai}$ $[\psi_{Ai} + \psi_{Aj} + \psi_{Bk}]$

ii) $\varepsilon_{Aj} < \varepsilon_{Ai}, \varepsilon_{Bk} < \varepsilon_{Ai}$ $[\psi_{Ai} + \psi_{Aj} - \psi_{Bk}]$ (A II-5)

iii) $\varepsilon_{Aj} < \varepsilon_{Ai}, \varepsilon_{Bk} > \varepsilon_{Ai}$ $[\psi_{Ai} - \psi_{Aj} + \psi_{Bk}]$

iv) $\varepsilon_{Aj} > \varepsilon_{Ai}, \varepsilon_{Bk} < \varepsilon_{Ai}$ $[\psi_{Ai} - \psi_{Aj} - \psi_{Bk}]$

since h'_{ik}s are all negative. In this Table the $+$ and $-$ signs in the bracket imply the in-phase and out-of-phase mixing of the subsequent orbital, with ψ_{Ai}, respectively. This presents the orbital-mixing rule to be obtained.

References

1) Born, M., Oppenheimer, J. R.: Ann. Physik **84**, 457 (1927).

2) Hartree, D. R.: Proc. Cambridge Phil. Soc. **24**, 89, 111 (1928).

3) Fock, V.: Z. Physik **61**, 126 (1930).

4) Kolos, W., Roothaan, C. C. J.: Rev. Mod. Phys. **32**, 219 (1960).

5) Kolos, W., Wolniewicz, L.: Rev. Mod. Phys. **35**, 473 (1963).

6) Buenker, R. J., Peyerimhoff, S. D., Whitten, J. L.: J. Chem. Phys. **46**, 2029 (1967).

7) Wahl, A. C., Bertoncini, P. J., Das, G., Gilbert, T. L.: Intern. J. Quant. Chem. **1**, 103 (1967).

8) Ritchie, C. D., King, H. F.: J. Chem. Phys. **47**, 564 (1967).

9) Pitzer, R. M.: J. Chem. Phys. **47**, 965 (1967).

10) Roothaan, C. C. J.: Rev. Mod. Phys. **23**, 69 (1951).

11) Slater, J. C.: Phys. Rev. **36**, 57 (1930).

12) Boys, S. F.: Proc. Roy. Soc. (London) A **200**, 542 (1950).

13a) Higuchi, J.: J. Chem. Phys. **22**, 1339 (1954).

13b) Karo, A. M.: J. Chem. Phys. **30**, 1241 (1959).

14) Löwdin, P.-O.: J. Chem. Phys. **18**, 365 (1950).

15) Clementi, E.: J. Chem. Phys. **36**, 33 (1962).

16) Hollister, C., Sinanoğlu, O.: J. Am. Chem. Soc. **88**, 13 (1966).

17) Snyder, L. C., Basch, H.: J. Am. Chem. Soc. **91**, 2189 (1969).

18) Roothaan, C. C. J.: Rev. Mod. Phys. **32**, 179 (1960).

19) Amos, A. T., Hall, G. G.: Proc. Roy. Soc. (London) A **263**, 483 (1961).

20) Das, G., Wahl, A. C.: J. Chem. Phys. **44**, 87 (1966). — Das, G.: J. Chem. Phys. **46**, 1568 (1967).

21) Pople, J. A.: Trans. Faraday Soc. **49**, 1375 (1953).

22) Kon, H.: Bull. Chem. Soc. Japan **28**, 275 (1955).

23) Pariser, R., Parr, R. G.: J. Chem. Phys. **21**, 466, 767 (1953).

24) Yonezawa, T., Yamaguchi, K., Kato, H.: Bull. Chem. Soc. Japan **40**, 536 (1967).

25) Kato, H., Konishi, H., Yamabe, H., Yonezawa, T.: Bull. Chem. Soc. Japan **40**, 2761 (1967).

26) Yonezawa, T., Nakatsuji, H., Kato, H.: J. Am. Chem. Soc. **90**, 1239 (1968).

27) Yonezawa, T., Kato, H., Kato, H.: Theoret. Chim. Acta **13**, 125 (1969). — Kato, H., Kato, H., Konishi, H., Yonezawa, T.: Bull. Chem. Soc. Japan **42**, 923 (1969).

28a) Pople, J. A., Santry, D. P., Segal, G. A.: J. Chem. Phys. **43**, S 129 (1965). — Pople, J. A., Segal, G. A.: J. Chem. Phys. **43**, S 136 (1965); **44**, 3289 (1966). — Pople, J. A., Beveridge, D. L.: Approximate molecular orbital theory. New York: McGraw-Hill Co., 1970.

28b) Pople, J. A., Beveridge, D. L., Dobosh, P. A.: J. Chem. Phys. **47**, 2026 (1967).

28c) Dewar, M. J. S., Haselbach, E.: J. Am. Chem. Soc. **92**, 590 (1970).

29) Katagiri, S., Sandorfy, C.: Theoret. Chim. Acta **4**, 203 (1966).

30) Imamura, A., Kodama, M., Tagashira, Y., Nagata, C.: J. Theoret. Biol. **10**, 356 (1966).

31) Hoffmann, R.: J. Chem. Phys. **39**, 1397 (1963).
32) Sandorfy, C.: Can. J. Chem. **33**, 1337 (1955).
33) Fukui, K., Kato, H., Yonezawa, T.: Bull. Chem. Soc. Japan **33**, 1197, 1201 (1960).
34) Hückel, E.: Z. Physik **60**, 423 (1930).
35a) O-ohata, K., Taketa, H., Huzinaga, S.: J. Phys. Soc. Japan **21**, 2306 (1966).
35b) Taketa, H., Huzinaga, S., O-ohata, K.: J. Phys. Soc. Japan **21**, 2313 (1966).
36) Hehre, W. J., Stewart, R. F., Pople, J. A.: J. Chem. Phys. **51**, 2657 (1969).
37) Van der Lugt, W. T. A. M., Ros, P.: Chem. Phys. Letters **4**, 389 (1969).
38) Clementi, E.: J. Chem. Phys. **46**, 3851 (1967).
39a) Dedieu, A., Veillard, A.: Chem. Phys. Letters **5**, 328 (1970).
39b) Duke, A. J., Bader, R. F. W.: Chem. Phys. Letters **10**, 631 (1971).
40) Lathan, W. A., Hehre, W. J., Pople, J. A.: J. Am. Chem. Soc. **93**, 808 (1971).
41) Clementi, E., Mehl, J., von Niessen, W.: J. Chem. Phys. **46**, 3851 (1967).
42) Schaefer, III, H. F.: The electronic structure of atoms and molecules. Addison-Wesley 1972.
43) Pauling, L., Wheland, G. W.: J. Chem. Phys. **1**, 362 (1933).
44) Brown, R. D.: Quart. Rev. **16**, 63 (1952).
45) Ri, T., Eyring, H.: J. Chem. Phys. **8**, 433 (1940).
46) Pullman, A., Pullman, B.: Experientia **2**, 364 (1946).
47) Dewar, M. J. S.: Trans. Faraday Soc. **42**, 764 (1946).
48a) Coulson, C. A.: Discussions Faraday Soc. **2**, 9 (1947); J. Chim. Phys. **45**, 243 (1948).
48b) Coulson, C. A., Longuet-Higgins, H. C.: Proc. Roy. Soc. (London) A **191**, 39; A **192**, 16 (1947).
49) Wheland, G. W.: J. Am. Chem. Soc. **64**, 900 (1942).
50) Dewar, M. J. S.: J. Am. Chem. Soc. **74**, 3357 (1952).
51a) Fukui, K., Yonezawa, T., Shingu, H.: J. Chem. Phys. **20**, 722 (1952).
51b) Fukui, K., Yonezawa, T., Nagata, C., Shingu, H.: J. Chem. Phys. **22**, 1433 (1954).
52) Fukui, K.: Molecular orbitals in chemistry, physics, and biology, p. 513 (ed. by P.-O. Löwdin and B. Pullman). New York: Academic Press 1964.
53) Nagakura, S., Tanaka, J.: J. Chem. Soc. Japan, Pure Chem. Sect. **75** 993 (1954).
54) Brown, R. D.: J. Chem. Soc. **1959**, 2232.
55) Mulliken, R. S.: J. Am. Chem. Soc. **74**, 811 (1952).
56) Mulliken, R. S.: Rec. Trav. Chim. **75**, 845 (1956).
57) Fukui, K., Yonezawa, T., Nagata, C.: Bull. Chem. Soc. Japan **27**, 423 (1954); J. Chem. Phys. **27**, 1247 (1957).
58) Fukui, K.: Modern quantum chemistry. Istanbul Lectures, Part 1, p. 49 (O. Sinanoğlu, ed.). New York: Academic Press 1965.
59) Woodward, R. B., Hoffmann, R.: J. Am. Chem. Soc. **87**, 395, 2511 (1965).
60) Fukui, K., Fujimoto, H.: Bull. Chem. Soc. Japan **41**, 1989 (1968).
61) Weinbaum, S.: J. Chem. Phys. **1**, 593 (1933).
62) Mulliken, R. S.: J. Chem. Phys. **36**, 3428 (1962).
63) Brillouin, L.: Actualites Sci. Ind. **1933**, 71; **1934**, 159.
64) Fukui, K., Fujimoto, H.: Bull. Chem. Soc. Japan **39**, 2116 (1966).
65) Mulliken, R. S.: J. Chim. Phys. **46**, 497 (1949).
66) Koopmans, T.: Physica **1**, 104 (1933).
67) Fukui, K., Fujimoto, H.: Bull. Chem. Soc. Japan **42**, 3399 (1969).
68) Rudenberg, K.: Rev. Mod. Phys. **34**, 326 (1962).
69) Hine, J.: J. Org. Chem. **31**, 1236 (1966) and references cited therein.
70) Fukui, K.: Bull. Chem. Soc. Japan **39**, 498 (1966).

References

71) Fukui, K.: Tetrahedron Letters **1965**, 2009.
72) Fukui, K., Fujimoto, H.: Mechanisms of molecular migrations, Vol. 2 (B. S. Thyagarajan, ed.). Wiley-Interscience 1969.
73) Streitwieser, Jr., A.: Molecular orbital theory for organic chemists. New York: John Wiley & Sons, Inc. 1961.
74) Dewar, M. J. S.: The molecular orbital theory of organic chemistry. New York: McGraw-Hill Book Co. 1969.
75) Fukui, K.: J. Chem. Soc. Japan (Ind. Chem. Sect.) **69**, 794 (1966).
76) Fukui, K., Kato, H., Yonezawa, T.: Bull. Chem. Soc. Japan **34**, 1112 (1961).
77) Fukui, K., Yonezawa, T., Nagata, C.: J. Chem. Phys. **27**, 1247 (1957).
78) Mulliken, R. S.: J. Chem. Phys. **23**, 1833, 1841, 2338, 2343 (1955).
79) Ruedenberg, K.: Rev. Mod. Phys. **32**, 335 (1960); **34**, 326 (1962).
80) Fukui, K., Fujimoto, H.: Tetrahedron Letters **1965**, 4303.
81) Brown, R. D.: J. Chem. Soc. **1959**, 2232.
82) Fukui, K., Yonezawa, T., Nagata, C.: J. Chem. Phys. **27**, 1247 (1957).
83) Hückel, E.: Z. Physik **76**, 628 (1932).
84) Moffitt, W.: Proc. Roy. Soc. (London) A **200**, 414 (1950).
85) Walsh, A. D.: J. Chem. Soc. **1953**, 2260, 2266, 2288.
86) Fukui, K., Imamura, A., Yonezawa, T., Nagata, C.: Bull. Chem. Soc. Japan **34**, 1076 (1961).
87) Huo, W. M.: J. Chem. Phys. **43**, 624 (1965).
88) Brion, H., Moser, C.: J. Chem. Phys. **32**, 1194 (1960).
89) Hazelrigg, Jr., M. J., Politzer, P.: J. Phys. Chem. **73**, 1008 (1969).
90) Shriver, D. F.: J. Am. Chem. Soc. **85**, 1405 (1963).
91) Bulgakov, N. N., Borisow, Y. A.: Kinetika i Kataliz **7**, 608 (1966).
92) Fukui, K., Morokuma, K.: Proc. Intern. Symposium Mol. Structure and Spectroscopy, Maruzen & Co. 1962, B — 121 — 1.
93) Clementi, E.: J. Chem. Phys. **46**, 4731 (1967).
94a) Al-Jaboury, M. T., Turner, D. W.: J. Chem. Soc. **1964**, 4438.
94b) Turner, D. W.: Tetrahedron Letters **1967**, 3419.
95) Bene, J. D., Jaffe, H. H.: J. Chem. Phys. **48**, 1807 (1968).
96) Kato, Hi., Kato, Ha., Konishi, H., Yonezawa, T.: Bull. Chem. Soc. Japan **42**, 923 (1969).
97) Tsubomura, H.: Bull. Chem. Soc. Japan **26**, 304 (1953).
98) Buenker, R. J., Peyerimhoff, S. D., Whitten, J. L.: J. Chem. Phys. **46**, 2029 (1967).
99) Pitzer, R. M.: J. Chem. Phys. **47**, 965 (1967).
100) Petke, J. D., Whitten, J. L.: J. Am. Chem. Soc. **90**, 3338 (1968).
100a) Lukina, M. Y.: Russ. Chem. Rev. **1962**, 419.
101) Pittman, Jr., C. U., Olah, G. A.: J. Am. Chem. Soc. **87**, 2998, 5123 (1965).
102) Closs, G. L., Klinger, H. B.: J. Am. Chem. Soc. **87**, 3265 (1965).
103) Deno, N. C., Richey, H. G., Liu, J. S., Lincoln, D. N., Turner, J. O.: J. Am. Chem. Soc. **87**, 4533 (1965).
104) Walsh, A. D.: Trans. Faraday Soc. **45**, 179 (1949).
105) Hoffmann, R.: Tetrahedron Letters **1965**, 3819
106) Palke, W. E., Lipscomb, W. N.: J. Am. Chem. Soc. **88**, 2384 (1966).
107a) Kato, H., Yamaguchi, K., Yonezawa, T., Fukui, K.: Bull. Chem. Soc. Japan **38**, 2144 (1965).
107b) Kato, H., Yamaguchi, K., Yonezawa, T.: Bull. Chem. Soc. Japan **39**, 1377 (1966).
108) Yamabe, S., Minato, T., Fujimoto, H., Fukui, K.: Theoret. Chim. Acta (Berl.) **32**, 187 (1974).

109) Fujimoto, H., Kato, S., Yamabe, S., Fukui, K.: J. Chem. Phys. **60**, 572 (1974).
110) Kato, S., Fujimoto, H., Yamabe, S., Fukui, K.: J. Am. Chem. Soc. **96**, 2024 (1974).
111) Wilke, G.: Angew. Chem. **72**, 581 (1960).
112) Chatt, J., Venanzi, L. M.: J. Chem. Soc. **1957**, 4735.
113) Goldstein, M. J.: J. Am. Chem. Soc. **89**, 6357 (1967).
114) Ohorodnyl, H. O., Santry, D. P.: J. Am. Chem. Soc. **91**, 4711 (1969).
115) Simmons, H. E., Fukunaga, T.: J. Am. Chem. Soc. **89**, 5208 (1967).
116) Goldstein, M., Hoffmann, R.: J. Am. Chem. Soc. **93**, 6193 (1971).
117) Kato, H., Yonezawa, T., Morokuma, K., Fukui, K.: Bull. Chem. Soc. Japan **37**, 1710 (1964).
118) Chan, A. C. H., Davidson, E. R.: J. Chem. Phys. **49**, 727 (1968).
119) Fujimoto, H., Fukui, K.: Tetrahedron Letters **1966**, 5551.
120) Kooyman, E. C., Vegter, G. C.: Tetrahedron **4**, 382 (1958).
121) Fujimoto, H., Fukui, K.: unpublished paper.
122) Fukui, K., Morokuma, K., Yonezawa, T.: Bull. Chem. Soc. Japan **34**, 1178 (1961).
123) Klopman, G., Hudson, R. F.: Theoret. Chim. Acta **8**, 165 (1967).
124) Hudson, R. F., Klopman, G.: Tetrahedron Letters **1967**, 1103.
125) Gloux, J., Guglielmi, M., Lemaire, H.: Mol. Phys. **17**, 425 (1969).
126) Klopman, G.: J. Am. Chem. Soc. **90**, 223 (1968).
127) Clar, E.: Polycyclic hydrocarbons, I and II. Academic Press 1964.
128) Salem, L.: J. Am. Chem. Soc. **90**, 543 (1968).
129) Salem, L.: J. Am. Chem. Soc. **90**, 553 (1968); Chem. Brit. **1969**, 449.
130) Clementi, E., Clementi, H., Davis, D. R.: J. Chem. Phys. **46**, 4725 (1967).
131) Paudler, W. W., Blewitt, H. L.: Org. Chem. **30**, 4081, 4085 (1965); **31**, 1295 (1966).
132) Streitwieser, Jr., A., Fahey, R. C.: J. Org. Chem. **27**, 2352 (1962).
133) Fukui, K.: Report Intern. Symp. Atom. Molec. Quant. Theory, Sanibel Island, Florida, Jan. 1964, p. 61.
134) Dewar, M. J. S.: Advan. Chem. Phys. **8**, 65 (1965).
135) Dewar, M. J. S.: The molecular orbital theory of organic chemistry, p. 331. New York: McGraw-Hill 1969.
136) Fujimoto, H.: unpublished paper.
137) Sagawa, S., Furukawa, J., Yamashita, S.: J. Chem. Soc. Japan (Ind. Chem. Sect.) **71**, 1897, 1900, 1909, 1913, 1919 (1968) (in Japanese).
138) Fukui, K.: Sigma molecular orbital theory (O. Sinanoğlu and K. B. Wiberg, ed.). The Yale University Press 1969.
139) Fukui, K., Kitagawa, Y., Hao, H., Fukui, K.: Bull. Chem. Soc. Japan **43**, 52 (1970).
140) Kato, H., Morokuma, K., Yonezawa, T., Fukui, K.: Bull. Chem. Soc. Japan **38**, 1749 (1965).
141) Fukui, K., Fujimoto, H.: Tetrahedron Letters **1965**, 4303.
142) Fujimoto, H.: unpublished paper.
143) Fukui, K., Hao, H., Fujimoto, H.: Bull. Chem. Soc. Japan **42**, 348 (1969).
144) Kwart, H., Takeshita, T., Nyce, J. L.: J. Am. Chem. Soc. **86**, 2606 (1964).
145) Goering, H. L., Nevitt, T. D., Silversmith, E. F.: J. Am. Chem. Soc. **77**, 4042 (1955).
146) Stork, G., White, W. N.: J. Am. Chem. Soc. **78**, 4609 (1956).
147) Fukui, K.: Nippon Kagaku Sen-i Kenkyusho Koenshu **23**, 75 (1966) (in Japanese).
148) Trevoy, L. W., Brown, W. C.: J. Am. Chem. Soc. **71**, 1675 (1949).

References

149) Hao, H., Fujimoto, H., Fukui, K.: Bull. Chem. Soc. Japan **42**, 1256 (1969).
150) Fujimoto, H., Oba, H., Fukui, K.: Nippon Kagaku Zasshi **90**, 1005 (1969) (in Japanese).
151) Kita, S., Fukui, K.: Nippon Kagaku Zasshi **88**, 996 (1967) (in Japanese).
152) Hoffmann, R., Woodward, R. B.: J. Am. Chem. Soc. **87**, 4388 (1965).
153) Herndorn, W. C., Hall, L. H.: Tetrahedron Letters **1967**, 3095 and others.
154) Fukui, K., Fujimoto, H.: Tetrahedron Letters **1966**, 251.
155) Marvell, E. N., Stephenson, J. L., Ong, J.: J. Am. Chem. Soc. **87**, 1267 (1965) and many other papers.
156) Doering, W. von E., Roth, W. R.: Tetrahedron **18**, 67 (1962); Angew. Chem. **75**, 27 (1963) and many other papers.
157) Schlatman, J. L. M. A., Pot, J., Havinga, E.: Rec. Trav. Chim. **83**, 1173 (1964).
158) Havinga, E., Schlatman, J. L. M. A.: Tetrahedron **15**, 146 (1961).
159) Fukui, K., Imamura, A., Yonezawa, T., Nagata, C.: Bull. Chem. Soc. Japan **33**, 1591 (1960).
160) Fukui, K., Fujimoto, H.: Bull. Chem. Soc. Japan **40**, 2018 (1967).
161) Fukui, K.: Tetrahedron Letters **1965**, 2427.
162) Bohm, B. A., Abell, P. I.: Chem. Rev. **62**, 599 (1962).
163) Goering, H. L., Nevitt, T. D., Silversmith, E. F.: J. Am. Chem. Soc. **77**, 4042 (1955).
164) Stork, G., White, W. N.: J. Am. Chem. Soc. **78**, 4609 (1956).
165) Schrage, K.: Tetrahedron Letters **1966**, 5795.
166) Berson, J. A.: Accounts Chem. Res. **1**, 152 (1968).
167) Lemal, D. M., McGregor, S. D.: J. Am. Chem. Soc. **88**, 1335 (1966).
168) Inagaki, S., Fukui, K.: Chem. Letters **1974**, 509.
169) Caple, R., Tan, H. W., Hsu, F. M.: J. Org. Chem. **33**, 1542 (1968).
170) Roth, W. R., Martin, M.: Ann. **702**, 1 (1967). — Gassman, R. G., Mansfield, K. T., Murphy, T. J.: J. Am. Chem. Soc. **91**, 1684 (1969).
171) Freeman, P. K., Raymond, F. A., Grostic, M. F.: J. Org. Chem. **32**, 24 (1967) and references cited therein.
172) Eliel, E. L., Ro, R. S.: J. Am. Chem. Soc. **79**, 5992 (1957). — Richer, J.-C.: J. Org. Chem. **30**, 324 (1965) and references cited therein.
173) Inagaki, S., Fukui, K.: Bull. Chem. Soc. Japan **45**, 824 (1972).
174) Houk, K. N., Strozier, R. W.: J. Am. Chem. Soc. **95**, 4094 (1973).
175) Criegie, R.: Angew. Chem. **74**, 703 (1962). — Wittig, G., Weinlich, J.: Chem. Ber. **98**, 471 (1965). — Hedeya, E., Miller, R. D., McNeil, D. W., D'Angelo, P. F., Schissel, P.: J. Am. Chem. Soc. **91**, 1875 (1969).
176) Houk, K. N.: J. Am. Chem. Soc. **95**, 4092 (1973).
177) Houk, K. N.: J. Am. Chem. Soc. **94**, 8953 (1972). — Houk, K. N., Sims, J., Duke, Jr., R. E., Strozier, R. W., George, J. K.: J. Am. Chem. Soc. **95**, 7287 (1973). — Houk, K. N., Sims, J., Watts, C. R., Luskus, L. J.: J. Am. Chem. Soc. **95**, 7301 (1973).
178) Minato, T., Yamabe, S., Inagaki, S., Fujimoto, H., Fukui, K.: Bull. Chem. Soc. Japan **47**, 1619 (1974).
179) Houk, K. N., Luskus, L. J., Bhacca, N. S.: J. Am. Chem. Soc. **92**, 6392 (1970).
180) Kita, S., Fukui, K.: Bull. Chem. Soc. Japan **42**, 66 (1969). — Fukui, K.: Accounts Chem. Res. **4**, 57 (1971). — Michl, J.: J. Am. Chem. Soc. **93**, 523 (1971); Mol. Photochem. **4**, 257 (1972). — Zimmerman, H. E., Epling, G. A.: J. Am. Chem. Soc. **94**, 3649 (1972).
181) Srinivasan, R., Boué, S.: Tetrahedron Letters **1970**, 203. — Srinivasan, R.: J. Am. Chem. Soc. **84**, 4141 (1962). — Boué, S., Srinivasan, R.: J. Am. Chem. Soc. **92**, 3226 (1970).

182) Baggiolini, E., Schaffner, K., Jeger, O.: Chem. Commun. **1969**, 1103. — Ipaktschi, J.: Tetrahedron Letters **1969**, 215. — Dauben, W. G., Kellogg, M. S., Seeman, J. I., Spitzer, W. A.: J. Am. Chem. Soc. **92**, 1786 (1970).
183) Inagaki, S., Fujimoto, H., Fukui, K.: J. Am. Chem. Soc., in press.
184) Inagaki, S., Minato, T., Yamabe, S., Fujimoto, H., Fukui, K.: Tetrahedron **30**, 2165 (1974) and references cited therein.
185) Inagaki, S., Yamabe, S., Fujimoto, H., Fukui, K.: Bull. Chem. Soc. Japan **45**, 3510 (1972).
186) Berson, J. A., Janusz, J. M.: J. Am. Chem. Soc. **96**, 5939 (1974).
187) Gould, E. S.: Mechanism and structure in organic chemistry, p. 295. New York: Henry Holt and Co. 1960.
188) Jacobson, S. E., Wojcicki, A.: J. Am. Chem. Soc. **95**, 6962 (1973).
189) Fukui, K., Inagaki, S.: J. Am. Chem. Soc., in press.
190) Merk, W., Pettit, R.: J. Am. Chem. Soc. **89**, 4788 (1967). — Hogeveen, H., Volger, H. C.: Chem. Commun. **1967**, 1133.
191) Halpern, J.: Accounts Chem. Res. **3**, 386 (1970).
192) Hogeveen, H., Volger, H. C.: J. Am. Chem. Soc. **89**, 2486 (1967).
193) Hogeveen, H., Volger, H. C.: Chem. Commun. **1967**, 1133.
194) Paquette, L. A.: J. Am. Chem. Soc. **92**, 5765 (1970); Accounts Chem. Res. **4**, 280 (1971).
195) Wristers, J., Brener, L., Pettit, R.: J. Am. Chem. Soc. **92**, 7499 (1970).
196) Ugo, R., Conti, F., Cenini, S., Mason, R., Robertson, G. B.: Chem. Commun. **1968**, 1498.
197) Chan, H. W. S.: Chem. Commun. **1970**, 1550.
198) Fukui, K., Ohkubo, K., Yamabe, T.: Bull. Chem. Soc. Japan **42**, 312 (1969))—Ohkubo, K., Yamabe, T., Fukui, K.: Bull. Chem. Soc. Japan **42**, 2220 (1969).
199) Nakamura, A., Otsuka, S.: J. Am. Chem. Soc. **94**, 1886 (1972).
200) Cowheard, F. G., von Rosenberg, J. L.: J. Am. Chem. Soc. **91**, 2157 (1969). — Roos, L., Orchin, M.: J. Am. Chem. Soc. **87**, 5502 (1965).
201) For instance see Calderon, N.: Accounts Chem. Res. **5**, 127 (1972) and references cited therein.
202) Yamamoto, T., Yamamoto, A., Ikeda, S.: J. Am. Chem. Soc. **93**, 3350 (1971).
203) For general treatment of such cases of MO crossing, see Bonacić-Koutecký, V., Koutecký, J.: Theoret. Chim. Acta (Berl.) **36**, 149, 163 (1975) and many papers cited therein.
204) Turro, N. J., Lechtken, P.: J. Am. Chem. Soc. **95**, 264 (1973). — Turro, N. J., Steinmetzer, H. C., Yekta, A.: J. Am. Chem. Soc. **95**, 6468 (1973). — Turro, N. J., Lechtken, P.: Pure Appl. Chem. **33**, 363 (1973). — Turro, N. J., Lechtken, P., Schore, N. E., Schuster, G., Steinmetzer, H. C., Yekta, A.: Accounts Chem. Res. **7**, 97 (1974).
205) Dewar, M. J. S., Kirschner, S.: J. Am. Chem. Soc. **96**, 7578 (1974). — cf. Yamaguchi, K., Fueno, T., Fukutome, H.: Chem. Phys. Letters **22**, 466 (1973).
206) Lechtken, P., Breslow, R., Schmidt, A. H., Turro, N. J.: J. Am. Chem. Soc. **95**, 3025 (1973).
207) Dewar, M. J. S., Kirschner, S., Kollmar, H. W.: J. Am. Chem. Soc. **96**, 7579 (1974).
208) McElroy, W. D., Seliger, E. H., White, E. H.: Photochem. Photobiol. **10**, 153 (1969). — White, E. H., Miano, J. D., Watkins, C. J., Breaux, E. J.: Angew. Chem. **86**, 292 (1974).
209) Goto, T., Kishi, Y.: Angew. Chem. Intern. Ed. Engl. **7**, 407 (1968).
210) Fukui, K.: J. Phys. Chem. **74**, 4161 (1970).
211) Bader, R. F. W.: Can. J. Chem. **40**, 1164 (1962).

125

References

212) Salem, L.: Chem. Brit. **1969**, 449.
213) Fukui, K.: The world of quantum chemistry, p. 113 (R. Daudel, B. Pullman, eds.). Dordrecht: Reidel Publ. Co. 1974.
214) Pearson, R. G.: J. Am. Chem. Soc. **91**, 4947 (1969); **94**, 8287 (1972); Theoret. Chim. Acta (Berl.) **16**, 107 (1970); Accounts Chem. Res. **4**, 152 (1971).
215) Fukui, K., Fujimoto, H., Yamabe, S.: J. Phys. Chem. **76**, 232 (1972).
216) Fujimoto, H., Yamabe, S., Fukui, K.: Bull. Chem. Soc. Japan **45**, 1566 (1972).
217) Fujimoto, H., Yamabe, S., Fukui, K.: Bull. Chem. Soc. Japan **45**, 2424 (1972).
218) Fujimoto, H., Yamabe, S., Minato, T., Fukui, K.: J. Am. Chem. Soc. **94**, 9205 (1972).
219) Fukui, K., Kato, S., Fujimoto, H.: J. Am. Chem. Soc. **97**, 1 (1975).
220) Herzberg, G.: Spectra of diatomic molecules, p. 322. Princeton: D. van Nostrand Co. 1950.
221) Eyring, H., Walter, J., Kimball, G. E.: Quantum chemistry, p. 205. New York: John Wiley and Sons 1944.
222) Woodward, R. B., Hoffmann, R.: The conversation of Orbital Symmetry. New York: Academic Press 1969.
223) Libit, L., Hoffmann, R.: J. Am. Chem. Soc. **96**, 1370 (1974). (In this paper all h'_{ij} and h_{kl} s are neglected).
224) Imamura, A.: A paper to be published in the J. Am. Chem. Soc. soon.

126

Author Index

Author Index

Venanzi, L. M. 46
Volger, H. C. 94, 95
von Rosenberg, J. L. 97

Wahl, A. C. 5, 6
Walsh, A. D. 40, 43, 45, 49, 50
Walter, J. 105
Watkins, C. J. 100
Watts, C. R. 77, 78
Weinbaum, S. 12
Weinlich, J. 76
Wheland, G. W. 8
White, E. H. 100
White, W. N. 58, 69
Whitten, J. L. 5, 45
Wilke, G. 46
Wittig, G. 76
Wojcicki, A. 92
Wolniewicz, L. 5

Woodward, R. B. 9, 33, 62, 64, 65, 67, 105
Wristers, J. 95

Yamabe, H. 6
Yamabe, S. 45, 78, 84, 104
Yamabe, T. 97
Yamaguchi, Ka. 6, 45
Yamaguchi, Ki. 100
Yamamoto, A. 98
Yamamoto, T. 98
Yamashita, S. 54
Yekta, A. 100
Yonezawa, T. 6, 8, 9, 22, 34, 36, 37, 39, 40, 41, 44, 45, 49, 50, 52, 53, 56, 57, 66

Zimmerman, H. E. 81

Subject Index

acrolein 76
activation energy 110
adamantane 55, 56
AgBF$_4$ 94
Alder rule 62
alkyl chlorosulfite 92
1,3-alkyl migration 91
allyl chloride 58
— rearrangement 58
alternant hydrocarbon 29, 39
aluminum hydride 45
t-amyl chloride 57
antarafacial migration 64
anti-interaction 69
antioxidant action 54
antisymmetrized product 2
axisymmetric process 76
azodicarboxylic ester 84
azulene 29

barbaralyl cation 92
BeH, BeH$_2$ 49
benzanthracene 51
benzene-silver cation complex 40
benzopentaphene 52
benzopyrene 51
BH$_3$CO 45
BH$_3$ dimer 45
B$_2$H$_6$ 45
bicycloheptene rearrangement 74, 76, 89
bicyclo [2.2.0] hexadiene 100
— [3.1.0] hexene 80
bicyclopentene 80
biphenylene 52
BNH$_6$ 45
bond eigenfunction 2
— exchange 113, 114
borazaphenanthrene 54
Born-Oppenheimer approximation 1, 10, 102

boron hydride 45, 46
Brillouin theorem 14
butadiene 71
4-t-butylcyclohexyl 50

carbon dioxide 84, 103
— monoxide 42
chair-boat selectivity 62
charge transfer 18, 21
— transfer force 8, 40
—-controlled reaction 33
—-transfer interaction 104
chemiluminescence 100
2-chloroethyl 50
2-exo-chloronorbornane 57
chloroparaffins 57
chlorosulfonyl isocyanate 84
N-chlorourethane 72
CI method 6, 12
Claisen rearrangement 62, 75
Cl$_2$CO 49
CNDO method 6, 43, 47
complete neglect of differential overlap 6
configuration interaction method 6, 12
conrotatory process 67, 68, 69, 106
Cope rearrangement 62, 63, 75, 90, 91
correlation diagram 105, 106, 107, 108
— error 5
Coulomb integral 23, 35, 37, 112
crystal-field interaction 33
cubane 94, 95
cyanide anion 43
cyclic azo compounds 75
— imines 76
cyclization 65, 66, 67
cyclobutadiene 77
cyclobutane 84, 95, 103
cyclobutene 94, 106
cyclohexanone 80
cyclooctadiene 46

131

Subject Index

cyclopropylcarbonium ion   45

deamination   74, 76
decarboxylation   89
delocalizability   9, 34, 36
delocalization approach   8
Dewar benzene   95, 100
dibenzoanthracene   51
dibenzochrysene   52
Diels-Alder reaction   26, 51, 60, 62, 69,
   70, 71, 76, 77
diene-dienophile interaction   60, 77
dimerization   98, 104
2,4-dimethyl-l-pentene   59
1,2-dioxetane   100
dioxetanone ring   101
1,3-dipolar addition   61, 77
dipolarophile   61
1,3-dipole   61
disproportionation   98
disrotatory process   67, 68, 106
dissociative adsorption   43
donor-acceptor interaction   20, 22, 31,
   37
Dr   36, 55, 56
dynamic orbital effect   118

E2 reaction   57
electrocyclic reaction   22
electron affinity   16, 51
— configuration   2, 11, 12
α, β-elimination   69
endo-exo selectivity   62, 76
ene reaction   90
enzymatic reaction   111
ethane   44
ethyl chloride   57
ethylene   85
— oxide   58
excited-configuration   11, 12, 14, 15, 85
exo-selectivity   79
extended Hückel method   6, 25, 30, 112

Fermi particle   2
fluoranthene   29, 30, 52, 53
formaldehyde   84
F_r 39
f_r 37, 38, 54, 55, 57

free valence   39
frontier electron theory   8, 9, 37
— MO   8, 21, 22, 26, 31, 32, 37, 39
— orbital density   8, 9, 24, 25, 27, 28,
   29, 30, 31, 34, 37, 39, 55, 56
—-controlled reaction   33
fulvene   92, 79

Gaussian type orbital   5
generalized coordinates   102
— frontier orbitals   33
GTO   5

Hamiltonian operator   1, 11, 117
Hartree-Fock method   3, 4, 5, 6, 11, 13,
   14
heterogeneous catalysis   43
— systems   111
hexatriene   67
highest occupied MO   8
   (see also "HOMO")
HNO_2 49
HOMO   8, 32, 67, 79, 84, 100, 103, 104
   and many other pages
homoaromaticity   47
HOMO-controlled 1,3-dipole   79
—-LUMO correlation   106
— — crossing   99, 105
— — separation   25, 108
homonuclear diatomic molecules   5
Hückel MO method   8, 25, 27, 30, 52,
   53
hybrid-based MO   6, 36, 55
hydrogen molecule   5, 12
1,3-hydrogen shift   97, 98

imidazopyridine   53, 54
INDO method   6
insertion   97
intermediate neglect of differential
   overlap   6
ionization potential   16, 43, 51

ketene   84
ketene immonium ion   84
Koopmans theorem   16

132

Reactivity and Structure

Concepts in Organic Chemistry
Editors: K. Hafner, J.-M. Lehn, C.W. Rees,
P.v. Rague Schleyer, B.M. Trost, R. Zahradnik

Volume 1: J. TSUJI
Organic Synthesis by Means of Transition Metal Complexes
A Systematic Approach
IX, 201 pages. 1975
ISBN 3-540-07227-6 Cloth DM 68,—
ISBN 0-387-07227-6 (North America) Cloth $27.90

Prices are subject to change without notice

This book is the first in a new series, Reactivity and Structure: Concepts in Organic Chemistry, designed to treat topical themes in organic chemistry in a critical manner. A high standard is assured by the composition of the editorial board, which consists of scientists of international repute.
This volume deals with the currently fashionable theme of complexes of transition-metal compounds. Not only are these intermediates becoming increasingly important in the synthesis of substances of scientific appeal, but they have already acquired great significance in large-scale chemical manufacturing. The new potentialities for synthesis are discussed with examples. The 618 references bear witness to the author's extensive coverage of the literature. This book is intended to stimulate organic chemists to undertake further research and to make coordination chemists aware of the unforeseen development of this research field.

Contents: Comparison of synthetic reactions by transition metal complexes with those by Grignard's reagent.
Formation of σ-bond involving transition metals.
Reactivities of σ-bonds involving transition metals.
Insertion reactions.
Liberation of organic compounds from the σ-bonded complexes.
Cyclization reactions.
Concluding remarks.

Springer-Verlag
Berlin
Heidelberg
New York

J. Falbe

Carbon Monoxide in Organic Synthesis

Translated from the German by Ch. R. Adams
21 figures. IX, 219 pages. 1970
ISBN 3-540-04814-6 Cloth DM 72,—
ISBN 0-387-04814-6
(North America) Cloth $19.90

The importance of carbon monoxide chemistry has increased rapidly in the last few years, both in scientific research and chemical processing. This necessitated the revision of the book, Synthesen mit Kohlenmonoxyd, published in German in 1967 as Vol. 10 in the series "Organische Chemie in Einzeldarstellungen". This covered Roelen's discovery of hydroformylation or oxo reaction, Reppe's carbonylation process, Koch's carboxylic acid synthesis and ring closure with carbon monoxide. The new edition includes latest research findings in these fields and, in particular, more space is given to the discussion of reaction mechanisms. The latest developments in industry are also mentioned and there have been numerous additions to the list of references.

This book will be an important reference source, both for the established expert and for those who wish to enter the fields of petrochemicals, organic chemicals and chemical engineering, particularly as results tend to be published in patents and thus remain outside the ken of a wide range of readers.

Contents: The Hydroformylation Reaction (Oxo Reaction/Roelen Reaction). — Metal Carbonyl Catalyzed Carbonylation (Reppe Reactions). — Carbonylation with Acid Catalysts (Koch Reaction). — Ring Closures with Carbon Monoxide. — Laboratory Preparations with Carbon Monoxide. — References. — Subject Index.

Prices are subject to change without notice

Springer-Verlag
Berlin
Heidelberg
New York